Cybersecurity

In today's hyperconnected world, digital threats lurk in every corner of cyberspace. This comprehensive guide equips you with essential knowledge to identify, mitigate, and counter cyber threats, navigating the increasingly dangerous digital landscape.

From cryptographic foundations to advanced persistent threats, you'll master the core concepts that transform novices into digital defenders. Each chapter tackles critical security domains, including online tracking defenses, ransomware identification, social engineering countermeasures, and OSINT techniques used by both defenders and attackers. Discover how to:

- Implement multi-layered protection against sophisticated malware
- Recognize and respond to psychological manipulation tactics
- Navigate the concealed territories of the Deep and Dark Web safely
- Deploy enterprise-grade network security architectures
- Leverage AI-driven threat detection to stay ahead of attackers.

Whether you're securing personal devices against tracking, protecting enterprise networks from nation-state actors, or building comprehensive defense strategies, this book is for working professionals, computing specialists, and IT enthusiasts, providing you with practical, actionable knowledge for today's security challenges.

Key Features:

- Bridges theoretical cybersecurity concepts with practical defense strategies
- Provides actionable knowledge on emerging threats
- Uses accessible language for both beginners and experienced practitioners
- Equips readers with a comprehensive understanding of the entire threat landscape
- Addresses both human and system vulnerabilities
- Delivers an insider perspective on threats and countermeasures
- Incorporates cutting-edge security approaches that prepare readers for the next generation of cybersecurity challenges

Cybersecurity
A Practical Introduction

Nihad A. Hassan and Rami Hijazi

CRC Press
Taylor & Francis Group
Boca Raton London New York

CRC Press is an imprint of the
Taylor & Francis Group, an **informa** business

A CHAPMAN & HALL BOOK

Designed cover image: Shutterstock

First edition published 2026
by CRC Press
2385 NW Executive Center Drive, Suite 320, Boca Raton, FL 33431

and by CRC Press
4 Park Square, Milton Park, Abingdon, Oxon, OX14 4RN

CRC Press is an imprint of Taylor & Francis Group, LLC

© 2026 Taylor & Francis Group, LLC

ISBN: 978-0-367-44214-9 (hbk)
ISBN: 978-0-367-44110-4 (pbk)
ISBN: 978-1-003-00827-9 (ebk)

DOI: 10.1201/9781003008279

Typeset in Palatino
by KnowledgeWorks Global Ltd.

To my mom Samiha, thank you for

everything… without you I'm nothing

Nihad A. Hassan

Contents

About the Authors

Nihad A. Hassan is an independent cyberse-curity consultant, digital forensics and cyber OSINT expert, online blogger, and author with over 15 years of experience in information security research. He has completed multiple technical security consulting engagements and authored six books and numerous articles on information security. Nihad is highly involved in security training, education, and motivation. For more information, go to www.osint.link.

Rami Hijazi is the general manager of MERICLER Inc., an education and corporate training firm in Toronto, Canada. Rami is an experienced IT professional who lectures on a wide array of topics, including object-oriented programming, Java, eCommerce, Agile devel-opment, database design, and data handling analysis. Rami also works as a consultant to Cyber Boundaries Inc., where he is involved in the design of encryption systems and wire-less networks, intrusion detection, and data breach tracking, as well as providing planning and development advice for IT departments concerning contingency planning.

1

Understanding Cybersecurity

Introduction

With the growth of digital technologies worldwide, the use of computerized systems to store and manipulate data has become prevalent in all work sectors. For instance, government, corporate, military, and health care organizations keep and process a large volume of digital data on computers and other digital devices. A substantial percentage of that information is sensitive (e.g., personal information, patient health information, intellectual property, military secrets, financial or other types) and must be protected from unauthorized access.

Organizations do not do business in isolation; they must transmit confidential information between separate networks and devices. This flow of digital information across computer networks – via intranet or the internet – will make it vulnerable to different threats. The risks to data are not merely limited when it is moving. For instance, information stored on computer systems is also susceptible to malicious attacks that aim to access, delete, or alter sensitive information; extort money by using ransomware or simply using Denial of Service (DoS) attacks to disturb normal network operations. Cybersecurity aims to mitigate and prevent all these types of cyberattacks.

Cybersecurity is the practice of protecting internet-connected devices (such as computers, servers, mobile & IoT devices, electronic systems, networks), computer applications, and the data stored within them from cyberattacks. Under this definition, cybersecurity contains all technologies, processes, and best practices to protect an organization from cyberattacks. This includes physical security, which involves protecting data centers and storage servers against unauthorized access, and logical security, which aim to protect data stored within computers storage and other digital media devices from all sorts of cyberattacks.

There are different definitions of the cybersecurity term proposed by official organizations; the following lists the main three:

- The *Committee on National Security Systems (CNSSI – 4009)* defines cybersecurity as "The ability to protect or defend the use of cyberspace from cyberattacks".

DOI: 10.1201/9781003008279-1

- *The National Institute of Standards and Technology (NIST)* defines cybersecurity as "The process of protecting information by preventing, detecting, and responding to attacks".
- *The International Organization for Standards (ISO/IEC 27032)* defines cybersecurity as the "Preservation of confidentiality, integrity, and availability of information in the Cyberspace".[1]

Cybersecurity threats and incidents are growing daily in cost, volume, and sophistication. They are projected to continue their massive growth over the coming years to private and public sectors. *Cybersecurity Ventures* predicted that cybercrime is expected to cost the world more than $10.05 trillion annually by 2025.[2] Cybercrime damage is not limited to the cost of destroying data (e.g., deleting it due to destructive malware or denying access to data through encrypting ransomware). For example, other losses count for the high cost, such as loss of productivity and disclosing sensitive information such as personal info and customers' medical and financial records. The cost also includes hiring digital forensics examiners and incidence response specialists to investigate data breaches, hiring specialized people in data recovery (to restore operations) without forgetting the reputation loss, which is considered the highest and needs years to fix.

In our first introductory chapter, we will thoroughly cover the term cybersecurity, define its types, and distinguish it from other IT security fields; we will discuss the different cyber threat types, threat actors, and common cybersecurity models. Cybersecurity frameworks that help organizations comply with various cybersecurity regulations will also get covered. We will also talk about cybercrime, its types, and motivation. However, before we begin our discussion, let us first define the virtual atmosphere where all these things happen, known as cyberspace.

Concept of Cyberspace

In a nutshell, cyberspace is the non-physical environment where all communication over computing networks occurs.

The term cyberspace was first coined in 1984 by a science fiction writer named *William Gibson* in his novel *Neuromancer* to describe his explication of a global network of interlinked computers. Under this definition, cyberspace includes all computing devices connected to the internet where users can send/receive emails, download/upload files, conduct online shopping and online banking, social media interactions (e.g., Facebook and Twitter), video conferencing, and chat. In addition, all collaboration tasks conducted by people residing in different geographical locations and connected through the internet are considered happening in this virtual – boundless – environment known as cyberspace.

Note! The cyberspace term has become a synonym for the Internet or the World Wide Web.

Now that we know what cyberspace is, we can clearly understand the term cybersecurity; it is the process of securing the information floating and all asset types (including the IT infrastructures and communications networks) contained in cyberspace.

Note! There are various accepted definitions of the term cyberspace created by multiple governments:

- *The United States, National Security Presidential Directive 54/ Homeland Security Presidential Directive 23 (2008)*: "The interdependent network of information technology infrastructures, and includes the Internet, telecommunications networks, computer systems, and embedded processors and controllers in critical industries".[3]
- *Canada's Cyber Security Strategy (2010)*: "Cyberspace is the electronic world created by interconnected networks of information technology and the information on those networks. It is a global common where more than 1.7 billion people are linked together to exchange ideas, services, and friendship".[4]
- *Germany, Cyber Security Strategy for Germany (2011)*: "The virtual space of all IT systems linked at data level on a global scale. The basis for cyberspace is the internet as a universal and publicly accessible connection and transport network which can be complemented and further expanded by any number of additional data networks. IT systems in an isolated virtual space are not part of cyberspace".[5]
- *The United Kingdom, The UK Cyber Security Strategy (2011)*: "Cyberspace is an interactive domain made up of digital networks that is used to store, modify and communicate information. It includes the internet, but also the other information systems that support our businesses, infrastructure and services".[6]

Cybersecurity Term Variations

Many people – including some IT professionals – use the terms "cybersecurity" and "Information Security" interchangeably when discussing the
security measures and steps needed to protect computer systems and information assets from cyberattacks. Although both terms are somehow related
in that context, they are not synonymous as many people think.

For instance, the main difference between cybersecurity and InfoSec is the
type of threat each one is concerned about. Cybersecurity focuses on outside
threats – those originating from the internet. In contrast, InfoSec focuses on
internal threats originating from within the subject organization (e.g., using
a USB device by a disgruntled employee to leak confidential information out
of the organization computers).

Note! What is the difference between Data and Information?
Data is raw facts (e.g., simple text and numbers) that must be processed first to become helpful in some context; data can only be considered information when it has a meaning. For example, "19800611"
is data; however, when it is interpreted as someone's data of birth
"1980/06/11", it becomes information.

InfoSec focuses on protecting data and information from unauthorized
access, accidental/intentional deletion, destruction, disclosure, or disruption to assure the confidentiality, integrity, and availability of these assets
to legitimate users. Under this definition, InfoSec is responsible for the
security of information in different environments, including the transfer of
confidential data outside cyberspace (e.g., encrypting backup hard drive/s
and moving them offsite fall under the responsibility of InfoSec) and even
protecting sensitive data when it is stored as physical files in the corporate
cabinet.

Cybersecurity deals with advanced cyberattacks coming from cyberspace,
such as ransomware, Advanced Persistent Threats (APT), cyber terrorists,
and black hat hackers trying to invade private networks and disrupt regular digital services. Cybersecurity is also concerned with the physical security of IT infrastructure and the security of online user accounts (e.g., social
media accounts like Facebook and Twitter, and email accounts) from data
breaches.

Cybersecurity spans its operations to protect computer communications
channels (both internet and intranet) and the data flow within them from all
types of malicious actors and threats, so these resources remain accessible
only to authorized users.

Note! We can consider InfoSec as a sub-branch of cybersecurity, although, as we already said, many people referred to cybersecurity as "Information Technology Security".

Cybersecurity Elements

As we saw, cybersecurity is a broad term used to describe all processes and procedures to secure computing systems (both hardware and software) and the data stored within them – or pass through – from cyberattacks. Based on this, cybersecurity can be found in all environments where IT systems need to be protected; the following are the most common cybersecurity subcategories.

Infrastructure Security (Physical Security)

Infrastructure security includes all infrastructures with access to the internet. For example, when an electricity grid has access to the internet, it will become vulnerable to cyberattacks. Securing physical IT infrastructure, especially public utilities systems (e.g., electricity, water, traffic lights, natural gas, and health centers), becomes necessary in today's digital age as modern socialites depend heavily on the internet to deliver public services. A good example of a cyberattack against a public utility is the attack against *Colonial Pipeline*, which took place in April 2021 and resulted in taking down the largest fuel pipeline in the US. This attack was found to be caused by a compromised password.

Network Security

Network security is an essential division within cybersecurity. It protects computer networks infrastructure and the data flowing across them from adversaries and malware threats. It is concerned with preventing malicious actors from sniffing traffic, delivering malicious software, or flooding servers with false traffic to disrupt normal business operations like DDoS attacks. The network security comprises two defense layers: one on the network gateways and the second within the networks. Network security controls include installing security appliances such as IPS/IDS and Firewalls, and governing users, applications, and systems access to protected resources.

Application Security

Application security prevents vulnerable applications from being installed in an organization's IT environment to avoid exploiting them by malicious actors. Application security should be well documented during the development and design phases, and after deployment to reveal any weakness or security vulnerability from being later exploited to facilitate intruders' exploits.

Endpoint Security

Endpoint security is concerned with protecting all endpoint computing devices (or end-user devices) on a network. Endpoint devices include desktops, servers, laptops, tablets, smartphones, and even Internet of Things (IoT) devices. Securing endpoint devices is a key to defending against cyberattacks. Such devices offer an entry point to hackers to penetrate an organization's network despite all security controls settled at network gates (firewall, IPS/IDS, and SIEM).

Since the start of the COVID-19 pandemic, more companies worldwide have allowed their employees to work remotely from home; securing employees' devices when used for work purposes also falls under this category.

Securing endpoints needs a combination of security policies and technologies (programs). Each one will be covered in a dedicated chapter later.

> Note! Some organizations include mobile security as a sub-branch of endpoint security. For instance, mobile security protects mobile computing devices (smartphones, tablets, PDA, IoT devices) and the wireless network they connect to from cyber threats. Some resources name mobile security as wireless security.

Data Security

Data security contains all processes, controls, and technologies used to secure data and protect it from unauthorized access; this also includes physical security controls to limit access to sensitive data storage locations. Data can be secured using different ways, such as:

- Encryption,
- Backup,
- Securely deleting data before dismissing old hardware storage units,
- Using a robust authentication mechanism, such as Two-Factor Authentication,
- Installing security solutions to stop malware on endpoint devices,
- Installing Network Detection and Response (NDR) solutions to protect data in hybrid environments (on-premises and cloud),
- Keeping everything up to date – such as OS and installed applications.

Cloud Security

Securing your data in the cloud should have equal importance to your on-premise data centers. However, when securing cloud data, you only deal with software solutions to monitor and protect the flow of information (up and down) from your local data centers to the cloud provider's servers.

Cloud security includes security policies and technological solutions to protect data stored within a cloud environment from theft, deletion, and data leakage. Such attacks can be originated from external sources or insider threats. Encryption is the primary technique used to protect cloud data from unauthorized access.

Cloud computing is evolving continually; there are different service models an organization – or a user – can choose from; the following lists the most popular three cloud service models.

1. Software as a Service (SaaS): In this model, the cloud providers host applications or services on their servers and make them accessible to users on a subscription basis via the internet. An example of this model is Microsoft Office 365, Salesforce, and Google Workspace.

2. Platform as a Service (PaaS): In this model, the cloud provider provides the platform, or operating system along with the required software development kits (such as Java SDK), in addition to providing storage and networking for the users. Users can deploy their applications and customize them if needed. Examples of this type are Amazon Web Services (AWS), Windows Azure, Apache Stratos, and Magento Commerce Cloud. PaaS model is commonly used for software creation projects.

3. Infrastructure as a Service (IaaS): In this service model, cloud users control cloud servers in terms of the operating system, applications, and networking; however, they can not control the underlying hardware running the cloud service.

These are the main cloud models; however, there are many more – or subsets – based on the previous models, such as data as a service (DaaS), Function as a service (FaaS), iPaaS (integration platform as a service), and security as a service (SECaaS).

Internet of Things (IoT)Security

IoT refers to all internet-capable devices that can exchange data and receive instructions via the internet. IoT devices contain many physical internet systems such as home appliances (washing machines, coffee makers, smart refrigerators, smart TV sets), healthcare devices (wearable sensors), security systems including alarms, Wi-Fi cameras, and even industrial equipment controlled remotely via the internet.

IoT security is concerned with safeguarding and securing IoT devices and associated networks from being abused for malicious purposes.

The fast pace at which people use IoT devices in their daily lives and their adoption in different work environments will make this sector very attractive to cybercriminals. The market size of IoT is enormous. According to *Statista*,[7] the global number of connected IoT devices is expected to reach 30.9 billion units by 2025.

End-user Cybersecurity Education

Cybersecurity awareness training is the most critical element in cybersecurity! Humans remain the first line of defense in any cybersecurity defense plan. For example, an unaware user can jeopardize the security of an organization's network by just clicking on a phishing link within a malicious email; this will effectively compromise the entire network despite all security solutions and controls already in place. A user can introduce threats to its work network using a variety of ways:

1. Install insecure programs from the internet,
2. Violating security guidelines when creating and storing passwords,
3. Using work email to register for free public online services,
4. Revealing important information (personal and work-related) on social media that can be used later by possible adversaries to formulate different social engineering attacks,
5. Inability to distinguish social engineering attacks, especially phishing attacks coming via email and social media messages.

Cybersecurity Models

An IT security model is a framework used by an organization to identify and develop security policies for a given context or setting (e.g., authentication policy). The security model's purpose is to outline the requirements needed to fulfill a specific security policy. The most common security models in use today are *CIA Triad* and the *Parkerian Hexad* model.

CIA Triad

This is the most widely accepted security model; the CIA comprises three fundamental principles (see Figure 1.1): Confidentiality, Integrity, and Availability. When planning any organization's security program, the security team needs to evaluate security threats and their potential impact on an organization's digital assets, such as data, systems, and applications,

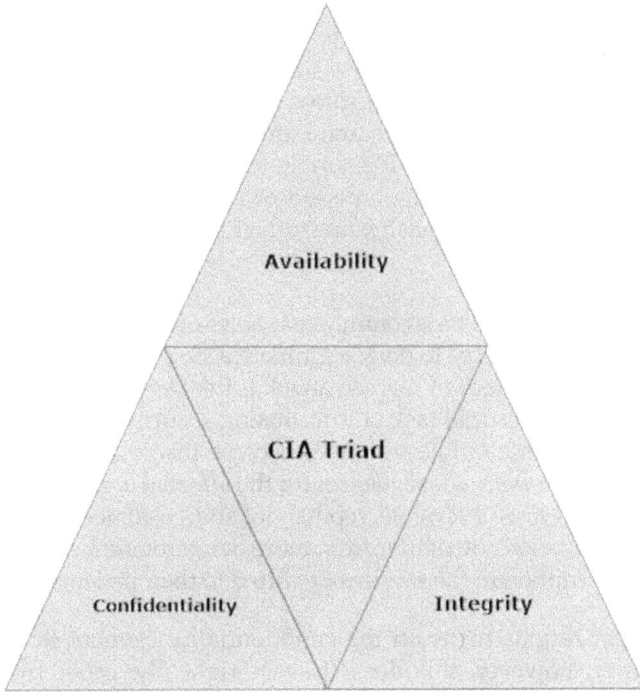

FIGURE 1.1
CIA triad.

regarding confidentiality, integrity, and availability. After the evaluation, the security team will draw the appropriate security policies and controls to mitigate these risks.

1. *Confidentiality*: When we say confidential data, we mean it must be secured against unauthorized access. Confidentiality can technically be achieved using access control lists and encryption to protect data transmitted through insecure mediums such as the internet. Data is commonly categorized according to its sensitivity level (For example, the level of damage it can bring if it falls into the wrong hands); after that, the corresponding security controls are implemented.

2. *Integrity*: Assure data is not deleted or modified by an unauthorized party. Keep in mind that this definition applies to legitimate users making unintentional mistakes when accessing this data, such as altering or deleting data accidentally, editing wrong files, and even corrupting files by introducing malware to the sensitive data storage media. A technical solution to ensure data integrity is using hashing; we should also consider keeping a current backup of all sensitive data.

Note! If you do not know what is meant by hash, check the "Hash Analysis" in Chapter 3. For now, remember that hashing works by converting a digital file (input) into a fixed string value (output); the resultant hash value is unique and cannot be generated again, using another file or piece of data. If a simple modification happened within the subject file (even deleting a period or changing the space between two words), the resultant hash value will change completely.

3. *Availability:* Assure data is continuously accessible to authorized parties. Some cyberattacks try to deny legitimate access to online resources; a well-defined instance of such an attack is the DoS attack. Maintaining availability is a crucial task of information security professionals. For example, suppose online resources become inaccessible. In that case, this will have severe consequences for the affected organization operations such as loss of revenue, reputation, and customers' trust. Health care organizations can suffer from more dangerous effects when there is an interruption in the services provided to their patients.

Many organizations focus on the confidentiality element to assure their data's security; however, this doesn't seem right; the other two elements should not get overlooked. Ultimately, the information is worthless if we cannot access it (Availability) and if it is not correct and reliable (Integrity).

Parkerian Hexad

This is a security model proposed by Donn B. Parker – who is considered a great mind in the information security field – built on the traditional CIA triad already discussed. The Parkerian model is not known as widely as the CIA triad model; however, it still provides more holistic view of today's complex IT landscape. As threats to information assets have developed in recent time, because of the increased growth of cloud services, the intense usage of the internet to transmit critical information such as health and financial data, besides the increased number of home-workers. Parker found the CIA triad was not comprehensive enough to cover all areas of today's modern information security environments, especially the human aspects of security, which address people's role to defend against cyber threats.

The *Parkerian* is composed of six elements, three from the CIA triad and the rest suggested by *Parker*:

1. *Confidentiality*
2. *Integrity*
3. *Availability*

4. *Possession or Control*: This element addresses the issue of breach of possession (the physical device that contains the data, such as a portable storage device, a laptop, tablet or a smartphone) in an information security context. For example, if you have sensitive info stored on a laptop and this laptop got stolen from you, then the person who acquires your laptop should not be able to view its contents although it is in his/her custody. How can this be achieved? For example, by using encryption to make stored data unreadable without a relevant decryption key. There is a breach of possession; however, data remain confidential because it is already encrypted.

5. *Authenticity*: This is all about assurance or trustees of the source of the received information online. Authenticity can be achieved technically by using a digital certificate, which is an electronic document used to identify an individual, a server, a company, or any other entity online.

6. *Utility*: This element points to how useful this data to us. For instance, we already gave an example about a stolen laptop containing sensitive info; if the laptop data was encrypted, this laptop would be of little utility to the attacker. On the other hand, if laptop data was not encrypted, this laptop would be of great utility to the attacker.

Cyber Threat Basic

Cyberthreat is any malicious activity that can lead to damage or interruption in computer networks or systems; it also includes any attempt to infiltrate IT systems to steal data or plant malware. Cyberthreats can be originated from either human or non-human sources. We can also differentiate it according to the origin of the physical location of the attacker, whether it is an "Insider Threat" or "Outsider threat".

Type of Cyberthreats

The criminals behind cyber threats are motivated by one of the following intents:

1. Monetary gain
2. Disruption of service
3. Espionage (corporate or state-sponsored)

Malicious actors use various attack techniques to execute their cyber threats. The following are the most common ones. (We will cover cyberattacks in a dedicated chapter later on.)

1. Malware
    1.1 Computer virus

  1.2 Worm

  1.3 Keylogger

  1.4 Rootkit

  1.5 Ransomware

  1.6 Botnets

  1.7 Trojan

  1.8 Spyware

  1.9 Drive-by Downloads

  1.10 And any computer program developed for malicious intents.

2. Social engineering attacks

3. Phishing

4. Malvertising

5. Exploit kit

6. DoS attack and Distributed Denial of Service (DDoS) attack

7. Advanced Persistent Threats

8. "Man in the Middle" (MitM) attack

9. Attacks on IoT Devices

The start of the COVID-19 pandemic was a turning point in the cyberthreat landscape. The massive shift of the workforce to become remote has expanded organizations' attack surface, making defending against cyberattacks more challenging and complex. On the other hand, the continual growth of cloud services along with the increased usage of IoT devices will increase the pressure on security teams to counter the ever-growing number of cyber attacks.

To have an idea about what the future hides for us in the cyberthreat landscape, check the following statistics:

1. According to IDC,[8] the global spending on cloud services will grow to reach $1.3 trillion by 2025.

2. Gartner[9] forecasts worldwide public cloud end-user spending to reach $362,263 million in 2022.

3. According to Statista,[10] the number of IoT-connected devices worldwide is projected to amount to 30.9 billion units by 2025.

4. A huge increase in the number of cyberattacks against cloud services was noticed since the start of the COVID-19 pandemic (January 2020). According to McAfee,[11] the estimated rise has reached 630%.

5. A Kaspersky research[12] found that more than 1.5 billion attacks have occurred against IoT devices in the first half of 2021.

The previous statistics show clearly that cybercriminals are shifting their attacks toward IoT devices and cloud services to cope with the new global technological trends. We have ascertained that this trend will remain in the future.

Where Are the Threats Coming From?

Cyberthreats can be originated from any of the following two sources: external or internal sources.

Outsider Threat

As its name implies, outsider cyber threats are external in origin and are the ones that most organizations focus their mitigation efforts on when creating their cyber defense plan. External threat actors work to invade private networks to gain unauthorized access and consequently to steal sensitive data – such as personal info, patients' health info, financial records, and trade secrets. Some external cyberattacks aim to disrupt normal network operations by launching a DDoS attack to cause revenue and reputation loss for the target organization (e.g., some hacktivists groups launch DDoS to promote their political and social agenda). Black hat hackers, organized criminals' groups, state-sponsored attacks, and hacktivists fall under this group. Conducting attacks from outside is difficult and costly; however, if succeed, it can cost the target organization millions of dollars of loss.

The most common attack vectors utilized by external threat actors are:

1. USB drop attack
2. Social Engineering attacks, such as phishing
3. DoS and distributed DoS attack
4. Hacking
5. Malware attacks – such as rootkits, backdoors, trojans, and ransomware attacks
6. Physical theft – such as stealing laptops or other storage media containing sensitive information

Organizations deploy various technological solutions to prevent external cyber threats, such as firewalls, web proxies, Intrusion detection and Intrusion prevention systems, Network Detection and Response (NDR), anti-malware solutions, DDoS attack mitigation, and external data loss prevention, to name a few.

Insider Threat

Insiders – without a doubt – are the most significant threat imposed on organizations. A malicious insider has all the necessary information to conduct his intrusion safely. Insiders already know where sensitive data is stored and how it is protected. They also know the sort of network configurations already in place and the type of applications already used. Besides, insiders can access sensitive areas and leak data without evidence of intrusion. We should note that not all insider threats are happening on purpose. For example, an unaware employee may store confidential business files on their laptop or USB device and accidentally lose them because of rubbery or other mistakes.

The *Verizon's 2021 Breach Investigations Report*[13] lists the main motivations behind insiders' threats as follows:

- Financial
- Espionage
- Fun
- Convenience
- Grudge
- Ideology

We should incorporate insiders' threats in every security defense plan. When the right security policies are set in place and enforced on all organization employees and external users who have legitimate access to the organization's IT system and network, and the right security tools are used correctly. Insider threats – including the one mentioned in our previous example – can be lowered to a great extent – for example, encrypting mobile devices' hard drives and setting a policy that prohibits storing sensitive files on personal storage devices. Preventing employees from attaching their mobile devices – including USB storage devices – to the corporate network can prevent serious data leaks caused by unaware employees.

Note! According to Panda Security,[14] insider threats have increased by 47% in the past two years. The average cost per incident because of employee or third-party contractor negligence was around $307.000 USD, while the cost of the incident due to criminal insider was over $347,130 USD, according to **secureworld**.[15]

Organizations commonly use Identity and Access Management (IAM) solutions to keep track of their employees' access credentials along with their access permission level. This ensures the right users have access to the right resources to do their job and avoid many threats associated with insiders. The expansion of organizations' supply chain networks introduces many third-party providers (contractors, subcontractors, service providers) to an organization network and IT system. Tracking all users, systems, applications, and other IT services access across an organization's IT environment becomes crucial to mitigate insider threats and avoid leaking sensitive data. This is what an IAM solution can do.

Another solution to stop internal data leakage is Data Loss Prevention (DLP); we will expand insider mitigations strategies in a later chapter.

Sources of Cyber Threats

It is equally important to identify a cyber threat to know who is behind the threat. Learning what motivates malicious actors to do their evil work is essential to plan mitigation strategies and protective measures. Cyberthreat techniques are evolving continuously; however, the sources behind these threats remain the same. The following lists the most common threat sources of cyberattacks.

State-Sponsored Attack

Also known as National cyber warfare, such attacks are conducted directly – or indirectly via proxy hacker groups – by a nation-state actor. These attacks are considered the most serious among all cyber threats' sources because of the resources available to fund the execution of the attacks and the possible damage expected. A nation-state attack's primary motivation is to steal national secrets (whether diplomatic, scientific, intellectual property theft, financial information, or military secrets). However, not all attacks are motivated by money; many attacks aim to disrupt services (such as public services like electricity, water supplies, internet services, traffic lights systems, and health information systems to name only a few) and bring damage to the target IT infrastructure and stored data. For example, many security experts think that the *NotPetya* ransomware – which has targeted major organizations in Ukraine in 2017 – is believed to be a Russian state-sponsored cyberattack. *NotPetya* is created to destroy data; its ultimate goal is not to generate profits from ransoms. Instead, it wants to sabotage and destroy data stored on target systems, leading to its classification as a cyber weapon.

There are many countries that are developing national cyber warfare programs and known to have interests in executing large-scale cyberattacks to disrupt the national public utilities infrastructure of their rivals. Four countries are famous for conducting state-sponsored cyberattacks: Russia, China, Iran, and North Korea.

Cyberterrorism

With the digitization of society, cyberspace becomes a new frontier for terrorist groups to conduct their terror activities.

As a concept, cyberterrorism includes any action carried out by terrorist groups – or by a single person – in cyberspace. It includes launching destructive attacks against computerized systems, spreading terror propaganda using the internet (e.g., using social media to recruit new fighters as with the case of ISIS), operating cyberspace to collect fund for terror groups, conducting reconnaissance activities online to carry out future attacks against physical targets and any act that can bring severe harm to the civilian population by using computer technology and the internet as a medium.

We should not confuse cyberterrorism with the cybercrime concept. For instance, cyberterrorism aims to do indistinguishable harm to many people for political or ideological reasons. Cyberterrorism act tends to be severe and seek to harm the civilian population without considering any ethical deterrents (e.g., using the internet to target connected critical medical devices may lead to the death of many patients whose lives depend on them). The same applies when a perpetrator/s tries to disrupt essential government infrastructure utilities such as water and electricity grid to bring harm and desperation to a vast population. A good example of a terror cyberattack is the attack in Oldsmar, a city of 15,000 people in the Tampa Bay area that happened in early 2021. Cyber attackers remotely accessed the water treatment plant and changed the levels of lye in the drinking water with the intent to poison the water supplies.[16]

We can differentiate between two types of cyberterrorists. The first one is using the internet as a communication channel. The second one uses it to launch cyberattacks – similar to black hat hackers, but of course, with much more harm and for different motives, as we already said.

Industrial Spies

Also known as Industrial Espionage, in this type, threat actors aim to steal intellectual property information and other trade secrets without the authorization of the owner of the information. In today's digital age, government organizations, research centers, universities, and business enterprises use IT to process and store their research data and intellectual property information. Threat actors know this and are continually developing sophisticated methods to launch attacks from cyberspace to illegally access this precious information.

Economic espionage can be conducted by different threat actors, such as nation-states, foreign corporations work under state influence, and proxy hacker groups. China's economic espionage is considered to have the most impact on the US economy; according to the *"Foreign Economic Espionage in Cyberspace"* report published by *National Counterintelligence and Security Center* in 2018,[17] the annual cost of China's economic espionage to the US economy can reach to $400 billion. The cost does not stop on the financial loss, as economic espionage can also damage the technological advancement achieved in many fields (e.g., scientific, space, and military) in favor of the perpetrator conducting the stealing act.

The wide adoption of Internet-of-Things (IoT) technology and the increased adoption of cloud services will introduce new vulnerabilities to IT networks; this will increase the attack surface, making cyber exploitation the preferred method of stealing proprietary information instead of using other traditional spying techniques like human recruitment and covert listening devices.

Black Hat Hackers

The internet has brought and will continue to bring considerable benefits to the global community; however, criminals also exploit anonymity and widespread internet usage worldwide to conduct their illegal activities safely. Hackers, or black hat hackers are the most well-known name for defining people using the internet to conduct illegal activities. Maybe this name is widely used by the public to refer to any activity conducted online with malicious intent. The act of conducting illegal activity online is known as cybercrime.

Cybercrime (also known as e-crime, high-tech crime, or digitized crime) includes all illegal activities committed in cyberspace; it can be broadly categorized according to the role of the computing device in the incident:

1. *The computer as a weapon*: Here, the computing device is used as a tool to commit a cybercrime such as launching ransomware campaigns, hacking into another computing device or network, cyberterrorism, sending phishing emails, and SPAM.
2. *The computer as a target*: The target, or victim, computing device suffer from a computer attack launched by one or more intruders (e.g., infecting with ransomware, becoming a target of DoS or DDoS attack).

Target of Cybercrime

Cybercrime can target any of the following entities:

1. *Cybercrime against individuals*: It includes email harassment, gaining unauthorized access to personal info (Identity Theft), cyberstalking, distributing pornography, and trafficking.
2. *Cybercrime against property*: It includes theft of confidential data such as bank account info, cyber vandalism, running phishing scams, and distributing malware.
3. *Cybercrime against the government*: Any cybercriminal activity conducted against the government is called cyberterrorism.

Cybercrime can be conducted by either an individual offender or a group of criminals.

Hacktivism

Hacktivism is the act of breaking into secure computer systems by an individual or a group of hacktivists to promote their political views or other social change agenda and draw people's attention to something the hacktivist believes in.

Hacktivist uses the same tools and techniques employed by black hat hackers to invade private networks; however, their ultimate goal is not to steal data for monetary gain or for espionage purposes, but to disrupt services and bring people's attention to a political or social cause. For example, a hacktivist group may break into the target organization's website and replace the homepage with an announcement (e.g., environment protection poster). A popular hacktivist attack uses a DDoS attack to bring down the target organization website or cease its online services.

Supporters of hacktivists argue that free speech laws protect hacktivism acts in disrupting online services. Although hacktivism attacks are not motivated by evil intentions; their expression method when attacking computer networks is considered illegal in many countries. For example, launching a DDoS attack is regarded as a federal crime in the United States. It is deemed to be illegal in many other countries such as Australia, the UK, and the EU countries.

There are many hacktivist groups, the most popular one: Anonymous and WikiLeaks.

Cybersecurity Frameworks

As the number of cyberattacks and data breaches continues to intensify, the demand to have a cybersecurity framework in place increases; such frameworks are suggested by leading cybersecurity organizations – and some of them are sponsored by government entities to enhance organizations defenses against cyberthreats. Some of these frameworks target one industry (e.g., financial or health sector), while others are more general and can be implemented across different sectors.

To define it, a cybersecurity framework is the set of measures, methodologies, procedures, and processes to align computer security policy, business operations, and technological defense measures to address cyber risks. For instance, most organizations – including those who work in the health sector – manipulate and store a large volume of personal data about their clients and customers. Exposing such info by offenders will make these organizations face lawsuits that will cost them huge fines without mentioning their reputation loss.

A cybersecurity framework helps an organization define its IT security needs and align those needs to its processes and technical defense measures.

Nowadays, conforming – to at least one – cybersecurity framework has become mandatory for doing business. Globalization, national and international data privacy regulations, the escalation of cyberterrorism and cybercrime activities in cyberspace, in addition to imposing compliance to such frameworks as a prerequisite for doing business with government institutions and large business organizations, all lead to making adopting a cybersecurity framework a must-have for any organization that wants to operate in today's digital age. Adopting a cybersecurity framework will also increase an organization's reputation against potential customers as a trusted partner who is immune – to a relatively good extent – against cyberattack.

> Note! *New Centrify* Survey that studies consumer attitude toward enterprise IT security finds that 66% of surveyed US Consumers are likely to cease doing business with a hacked organization.[18]

According to *Tenable's* survey titled "Trends in Security Framework Adoption".[19] It finds that 84% of organizations in the USA already leverage one security framework in their organization, while finding 44% are using more than one framework. This survey lists the most general cybersecurity framework already in use:

1. Payment Card Industry Data Security Standard (PCI DSS) (47%) (https://www.pcisecuritystandards.org)
2. ISO/IEC 27001/27002 (ISO) (35%) (https://www.iso.org/isoiec-27001-information-security.html)
3. Center for Internet Security Critical Security Controls (CIS) (32%) (https://www.cisecurity.org/controls)
4. National Institute of Standards and Technology (NIST) Framework for Improving Critical Infrastructure Cybersecurity (NIST CSF) (29%) (https://doi.org/10.6028/NIST.CSWP.04162018)

Why Cybersecurity Is Considered a Good Career?

Experienced cybersecurity professionals are in high demand; as the world continues to digitalize, cybersecurity professionals' need to protect digital assets increases. The demand for qualified cybersecurity professionals is growing much faster than any other industry. Nowadays, there is a significant shortage of cybersecurity professionals worldwide. The job market cannot keep pace with the continual escalation of cybercrime. Cybersecurity

Ventures predicts there will be 3.5 million unfilled cybersecurity positions in 2021, and the same number of openings will remain until the year 2025.[20]

In addition to the low unemployment rate, there are different reasons for making cybersecurity an attractive profession:

1. Good salary: All cybersecurity-related jobs pay very well, especially for the senior level positions who gets the highest salaries, which reach up to $400.000 annually. To have an idea about the different salary scales for different cybersecurity roles, Cyberseek[21] has developed an interactive career pathway map that displays key cybersecurity jobs, common transition opportunities between them, and information about the salaries, credentials, and skillsets associated with each role (see https://www.cyberseek.org/pathway.html).

2. Can work in any industry: An interesting thing about working in cybersecurity is your ability to work in any industry. For instance, medical centers, car manufacturers, banks, financial organizations, universities, retail, media companies, and any organization (both private and public sectors) that utilizes digital technology in its operations need to hire cybersecurity experts and consultants to protect its computer networks and sensitive data.

3. Huge growth potential: Cybersecurity field spans over wide domains and provides its learners with great growth opportunities. For instance, as digital technology continues to advance, the adoption of cloud services, AI, ML, and IoT devices is witnessing significant growth. To become competent with the latest technologies, a cybersecurity professional must master different technologies, such as computer networking, penetration testing, understand cloud architecture, be familiar with the different data privacy regulations besides mastering other non-technological areas such as creating IT security plans and polices besides mastering using a plethora of solutions to defend against cyberattacks.

4. A challenging profession: Unlike other professions, Cybersecurity profession requires continual skills update to remain in-line with the latest cyber threats. As technology becomes integrated in all our life aspects, the need to seek advice and services from cybersecurity professionals increase. The emergence of new threats and the need to deploy and use new technology and tools to counter those threats makes cybersecurity domain very challenging and full of excitement.

The previous reasons clearly show that planning a career in this field is worth the effort, and arming yourself with cybersecurity certifications will boost your career and make you stand out among your peers.

Cybersecurity Certification

Today, there are many cybersecurity certification programs; if you are confused and do not know where to start, the following lists the most sought-after cybersecurity certifications in North America – according to the "2021 IT Skills and Salary" report released by Global Knowledge.[22]

1. Google Certified Professional Cloud Architect (https://cloud.google.com/certification/cloud-architect)
2. Google Certified Professional Data Engineer (https://cloud.google.com/certification/data-engineer)
3. AWS Certified Solutions Architect – Associate (https://aws.amazon.com/certification/certified-solutions-architect-associate)
4. CRISC – Certified in Risk and Information Systems Control (http://www.isaca.org/Certification/CRISC-Certified-in-Risk-and-Information-Systems-Control/Pages/default.aspx)
5. CISSP – Certified Information Systems (https://www.isc2.org/Certifications/CISSP)

CISM – Certified Information Security Manager (http://www.isaca.org/Certification/CISM-Certified-Information-Security-Manager/Maintain-Your-CISM/Pages/default.aspx)According to the report, cloud computing and cybersecurity certifications are still in high demand globally, and make up the five highest-paying certifications in the US and Canada. The future seems bright for cloud technology because of its accelerated adoption worldwide. The report has an outlook toward the future and lists the most pursued certifications by IT professionals worldwide according to the current investment areas in IT:

1. CISSP – Certified Information Systems Security Professional
2. AWS Certified Solutions Architect – Associate
3. AWS Certified Cloud Practitioner
4. CISM – Certified Information Security Manager
5. Google Cloud Professional Cloud Architect
6. AWS Certified Solutions Architect – Professional
7. Microsoft Certified: Azure Fundamentals
8. CRISC – Certified in Risk and Information Systems Control
9. Microsoft Certified: Azure Administrator Associate
10. Microsoft Certified: Azure Solutions Architect Expert

Summary

There are fantastic benefits of adopting technology in all life and work areas. Technology allows faster information sharing, reduces costs, and allows work collaborations between people living in separate physical locations, without forgetting the significant benefits technology brought to our personal lives. However, this comes with a cost; the internet offers a unique opportunity for offenders to use it anonymously to launch cyberattacks and use it as a secure communication channel with minimal possibility of being discovered.

Cybersecurity is the name used to describe processes and technical procedures to safeguard IT systems from cyber threats. Cybersecurity has become integrated into all business sectors where IT systems need to be protected. I today's digital age, technology is not limited to using individual computing devices and their applications. Organizations of all sizes and industries use technology in their daily operations, from email to data storage and reaching to IoT and mobile computing. Technology has become everywhere, and without it, most organizations can not do their regular daily work.

Threats can be classified in cyberspace in many ways. The best method is to classify them according to their origin or source, insider or outsider threats. The motivation for conducting cyberattacks differs according to each threat actor's intent and ranges from financial gain to espionage and ending with causing damage as with the terror attacks.

Adopting a cybersecurity framework becomes mandatory for all organizations of any size or industry. Compliance with a cybersecurity framework allows an organization to identify cyber threats correctly and adopt the necessary security policies and technical defenses to mitigate them.

This chapter introduces the big concept of cybersecurity; in the next two chapters, we will continue our discussion to cover other important info that any cybersecurity professional must understand, like internet layers, the darknet, and the basics of cryptography systems.

Notes

1 ISO, "Guidelines for Cybersecurity", Accessed 2025-04-02. https://www.iso.org/obp/ui/#iso:std:iso-iec:27032:en
2 Cybersecurity ventures, "Cybercrime to Cost the World $10.5 Trillion Annually By 2025", Accessed 2025-04-02. https://cybersecurityventures.com/cybercrime-damage-costs-10-trillion-by-2025
3 NIST, "Computer Security Resource Center", Accessed 2025-04-02. https://csrc.nist.gov/glossary/term/cyberspace
4 Government of Canada Publications, "Canada's Cyber Security Strategy", Accessed 2025-04-02. http://publications.gc.ca/collections/collection_2010/sp-ps/PS4-102-2010-eng.pdf
5 CIO, "Federal Ministry of the Interior: Cyber Security Strategy for Germany", Accessed 2025-04-01. http://www.cio.bund.de/SharedDocs/Publikationen/DE/Strategis

6   Gov.UK, "The UK Cyber Security Strategy", Accessed 2025-04-02. https://assets. publishing.service.gov.uk/government/uploads/system/uploads/attachment_ data/file/60961/uk-cyber-security-strategy-final.pdf
7   Statista, "Internet of Things (IoT) and Non-IoT Active Device Connections Worldwide from 2010 to 2025", Accessed 2025-04-02. https://www.statista.com/ statistics/1101442/iot-number-of-connected-devices-worldwide
8   Businesswire, "IDC Forecasts Worldwide "Whole Cloud" Spending to Reach $1.3 Trillion by 2025", Accessed 2025-04-01. https://www.businesswire.com/news/ home/20210914005759/en/IDC-Forecasts-Worldwide-Whole-Cloud-Spending-to-Reach-%241.3-Trillion-by-2025
9   Gartner, "Gartner Forecasts Worldwide Public Cloud End-User Spending to Reach Nearly $500 Billion in 2022", Accessed 2025-04-01. https://www.gartner. com/en/newsroom/press-releases/2020-11-17-gartner-forecasts-worldwide-public-cloud-end-user-spending-to-grow-18-percent-in-2021
10 Statista, "Internet of Things (IoT) and Non-IoT Active Device Connections Worldwide from 2010 to 2025", Accessed 2025-04-01. https://www.statista.com/ statistics/1101442/iot-number-of-connected-devices-worldwide/
11 Compliancy-group, "Cyber Attacks on Cloud Services Rise 630%", Accessed 2025-04-01. https://compliancy-group.com/cyber-attacks-on-cloud-services-rise-630/
12 Digit, "IoT Devices See More Than 1.5bn Cyberattacks so Far This Year", https:// www.digit.fyi/iot-security-kaspersky-research-attacks
13 Verizon, "Data Breach Report", Accessed 2025-04-02. https://enterprise.verizon. com/resources/reports/2021/2021-data-breach-investigations-report.pdf
14 Techjury, "22 Insider Threat Statistics to Look Out For in 2024", Accessed 2025-04-02. https://techjury.net/blog/insider-threat-statistics/
15 Secureworld, "Ponemon: Data Theft by Criminal Insiders Has a High Price", Accessed 2025-04-02. https://www.secureworld.io/industry-news/data-theft-by-criminal-insiders-has-a-high-price
16 Nytimes, "Dangerous Stuff': Hackers Tried to Poison Water Supply of Florida Town", Accessed 2025-04-02. https://www.nytimes.com/2021/02/08/us/olds-mar-florida-water-supply-hack.htmlt
17 Dni, "Foreign Economic Espionage in Cyberspace", Accessed 2025-04-02. https:// www.dni.gov/files/NCSC/documents/news/20180724-economic-espionage-pub.pdf
18 Businesswire, "New Centrify Survey Finds 66 Percent of U.S. Consumers Are Likely to Stop Doing Business with a Hacked Organization", Accessed 2025-04-02. https://www.businesswire.com/news/home/20160608005485/en/Centrify-Survey-Finds-66-Percent-U.S.-Consumers
19 Tenable, "Survey Report: Trends in Security Framework Adoption", Accessed 2025-04-02. https://www.tenable.com/whitepapers/trends-in-security-framework-adoption
20 Cybersecurityventures, "Cybersecurity Jobs Report 2018-2021", Accessed 2025-04-02. https://cybersecurityventures.com/jobs/
21 Cyberseek, "Cybersecurity Career Pathway", Accessed 2025-04-02. https://www. cyberseek.org/pathway.html
22 Globalknowledge, "2021 IT Skills and Salary Report", Accessed 2025-04-02. https://www.globalknowledge.com/us-en/content/salary-report/it-skills-and-salary-report

2

Cyber Anonymity and Web Layers

Introduction

The proliferation of internet technologies worldwide has been accompanied by an equal increase in cybercriminal activities. The anonymous nature of cyberspace allows cybercriminals to disguise their true identity when conducting illegal activities; this increases criminal motivation and becomes an issue of broad concern.

On the other side, we cannot consider cyber anonymity as a direct facilitator of cybercrime. For instance, whistleblowers, journalists, and political activists living in countries controlled by oppressive regimes need to communicate and post their ideas in cyberspace anonymously. In addition to those, most internet users are not criminals, and they have the right to protect their browsing activities and prevent outsiders from tracking their web surfing activities. Privacy and anonymity are fundamental concepts in cybersecurity and affect us in our daily lives, so it is vital to distinguish between the two terms. Privacy means keeping your online activities private; although these activities can be visible to everyone, their contents should remain confidential. For example, when you send an encrypted email to a friend, only your friend should be able to read it; however, the existence of this communication between you and him is not secret or anonymous.

On the other hand, anonymity is about keeping your true identity secret. However, other parties can know what you do (e.g., post something to a discussion form under a false name). In other words, anonymity means no one can attribute your online actions to your true identity.

Cybersecurity is concerned with attacks emerging from cyberspace; this chapter will introduce the web layers and see how to search within them efficiently, focusing on the Deep and Darknet layers because of their importance to cybersecurity operations. Becoming anonymous in cyberspace is an essential skill that any cybersecurity professional should master. However, before we begin, we should understand how internet users can be tracked online and explore the best methods to prevent external actors from tracking internet users' traces.

DOI: 10.1201/9781003008279-2

Note! Online privacy needs a book on its own; the author advises you to grab his dedicated book about digital privacy and cyber anonymity titled *"Digital Privacy and Security Using Windows"* published by Apress 2017, for full coverage of the topic.

Web Tracking

The internet is considered a non-secure environment where different entities track users' browsing activities for varying reasons. It is essential for cybersecurity professionals to understand how web tracking works for the following reasons:

- To browse the web securely and privately.
- To learn to cover their true digital identity – and become fully anonymous – when doing online searches.
- To access the darknet networks, such as The Onion Router (TOR), anonymously.
- Understanding how web tracking works is essential when conducting digital forensics investigations and searching for incriminating evidence online.

This section will define the term web tracking, see who is interested in conducting such action, learn how web tracking works technically, and advise countermeasures against it.

Parties Interested in Tracking Internet Users

Different parties are interested in tracking internet users' activities; this section will list them and see what motivates them.

Advertising Companies

Online Advertisements (Ads) are widespread in cyberspace. Whenever you use a free online service (e.g., Free email service, watch YouTube videos, conduct a Google search, or surf Facebook) or install a free mobile app, there is a high probability that this service/App is funding itself by displaying promotional advertisements. Online advertisements are what make most of the internet free!

Note! In the third quarter of 2021, Facebook's total advertising revenue reached 28.2 billion U.S. dollars.[1]

We can distinguish between two entities when talking about online Ads, the *Advertiser* who pays money to display its Ads, and the *publisher* who gets paid for displaying these ads on its internet channel (e.g., Website or Mobile Application). Keep in mind that an advertiser can also own the medium used to display the Ads making it plays two roles.

Online advertisements are usually tailored according to each user's preferences. To achieve a high return from these Ads, Advertisers – and the tracking companies they hire – use different techniques to track and profile internet users to target them with customized Ads that match their usage patterns and location. Online tracking techniques will be covered thoroughly in the coming section.

Law Enforcement & Security Services Agencies

Law enforcement tracks persons of interest online activities to solve crimes and gather valuable intelligence about any entity online. As the world continues to digitize, different kinds of information become available about any internet user, such as previous purchases, location, history, personal info, list of friends and work colleagues in addition to other personal facts like online habits, questions asked to Google, the articles a user read, political views and even user health status.

While monitoring people's online activities requires a search warrant in most democratic countries, this is not always applied. For instance, police and security services agencies were found –in many cases – to gather intelligence on a mass scale about citizens without acquiring such legal consent.[2]

Note! Open Source Intelligence (OSINT)

OSINT is an acronym that points to all the information that can be gathered from public sources freely (Internet, books, magazines, public databases, corporate and academic publications, newspapers, TV, radio, to name a few) and legally without breaching any privacy or copyright laws. Nowadays, OSINT tools and techniques are used extensively by law enforcement and security services to gather valuable intelligence about any entity online, in addition to using it by enterprises for competitor analysis. OSINT will be covered in a dedicated chapter later on in this book.

Web Analytics

Web analytics services – such as Google Analytics – track and record visitor interactions with a website for marketing and statistical purposes. Information gathered through web analytics includes time spent on a page, exit and entrance pages, visitor behaviors, popular pages, countries of origin, in addition to device and browser types.

To use such analytical services, website owners should place a tracking code on each web page they want to track. Nearly all websites use web analytics to optimize the user experience and use the collected data in targeted marketing.

Hacking and Malware

Malicious actors can exploit some online tracking techniques to identify vulnerable targets online. For instance, a malicious actor can set up a website with a fingerprinting script within it to identify visitors' device technical info (e.g., browser and OS type, installed browser add-ons). Now attackers compile a list of targets that are still using vulnerable software and exploit them using different attack techniques.

A new cyber-attack based on web browser spoofing is used to gain unauthorized access to internet users' accounts. The attack works by controlling a compromised website containing a fingerprinting script. When the unaware user visits the attacker's website, the script collects all technical information about the user device (hardware and software configuration settings). The collected information is then used to spoof the user device's digital fingerprint by creating a complete replication of the user computing device software and hardware stack using a virtual machine. This allows attackers to access victims' online accounts that rely on device fingerprinting recognition (see Figure 2.1).

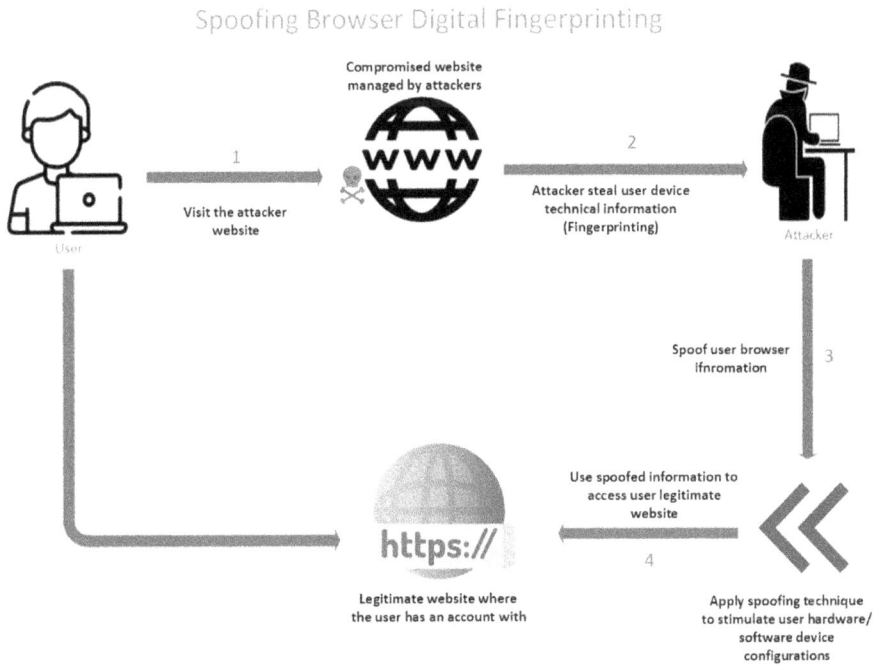

FIGURE 2.1
A new attack allows hackers to spoof victim digital browser fingerprint to fool monitoring system of other websites to gain unauthorized access to victim accounts.

Difference Between First-party & Third-party Web Tracking

When talking about web tracking, we will usually hear the terms first-party and third-party web tracking; let us explain the difference between both using a simple example.

Suppose you go to the CNN website and read an article; the CNN website will record this visit, know which article you read, and the amount of time you spent reading it, among other parameters. In this example, the CNN website is a first-party tracker.

Now let us assume CNN is hiring another company (e.g., Double Click, TagMan, Ominture, DC Storm, Site Tagger) to manage the tracking technologies and customize advertisements shown on its website. In that case, the tracking company is named a third-party tracker.

Third-party trackers impose a privacy risk because of their ability to record/track your online activities across different websites without your knowledge. Third-party trackers can also invite more third-party trackers to the first-party webpage, thus making your internet browsing history exposed against other trackers.

> Note! Many people think web tracking is anonymous as the tracker (whether it is a First-party or Third-party) cannot link your browsing history with your real identity. However, this is not accurate, as we will see in "Examples of Online Tracking" later.

Online Tracking Mechanisms

Online tracking can be defined as recording and storing internet browsing history (and even online behavior, preferences, and even every click!) of internet users across different websites. Most trackers are concerned about profiling internet users to create a complete profile for each internet user's online activities. To achieve this, online trackers should link a user's browsing history with its real identity using an identifier. This identifier is similar to a person's fingerprint as it can distinguish a particular user computing device among billions of connected devices.

Next, we will see how popular online tracking methods work technically.

IP Address

Whenever you connect to the internet, your device – and current geographical location – can be identified by a unique number called IP address or "Internet Protocol". An IP address is considered unique as no two devices can have the same IP address on the same IP network (e.g., internet). This makes an IP address the first option for online trackers when tracking internet users browsing activities.

TABLE 2.1

IPv4 & IPv6 Format

Example of IPv4	192.168.1.6
Example of IPv6	0:0:0:0:0:FFFF:C0A8:0106

There are two versions of the Internet Protocol, IPv4, which is the old version and uses 32-bit number, and IPv6, which is the new standard and uses 128-bit to create a single IP address (see Table 2.1). The majority of personal computing devices worldwide still use IPv4.

An IP address can be either static or dynamic. Static IP addresses (also known as dedicated or fixed IP addresses) are assigned by the Internet Service Provider (ISP) and do not change when the user initiates a new internet connection or reboots their computing device or router. However, a user needs to configure its computer network settings manually to use it. This type is rarely used for personal devices, as we commonly see on email servers, network printers, and file servers.

Most IP addresses on the web are dynamic; a dynamic address will change each time a user connects to the internet and is assigned using Dynamic Host Configuration Protocol (DHCP). A DHCP is a service running on networks (Routers or dedicated DHCP servers) that serves as a coordinator for assigning IP addresses for all devices on the same network.

To see your current IP address and whether it is static or dynamic under Windows OS, do the following:

Open a command-line prompt and type **ipconfig/all** (see Figure 2.2).

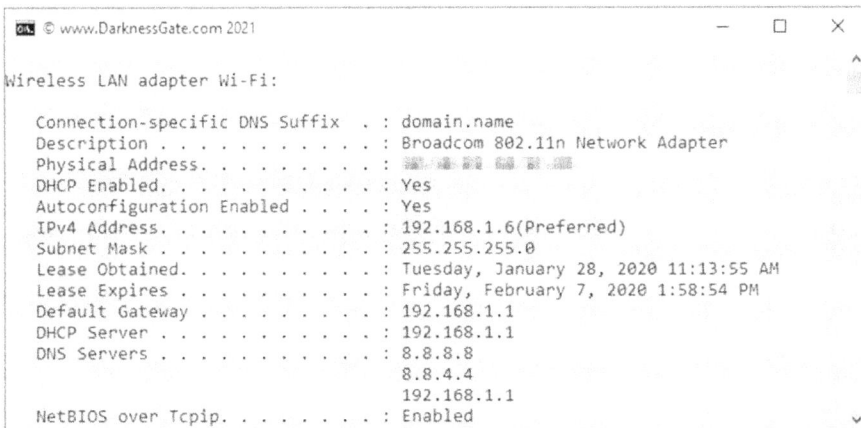

FIGURE 2.2
When DHCP value of your current active connection is set to "Enabled", this means the client is using a dynamic IP address.

Although IP address is considered an essential parameter in tracking online users, it cannot be considered alone enough to uniquely track internet users' devices. For instance, real IP addresses can be concealed using a Virtual Private Network (VPN) connection or anonymity networks such as the TOR network. Many internet users also sit behind a Network Address Translation (NAT) router that displays one public IP address and hides all private IP addresses of all devices connecting to the internet.

> Note! Although a dynamic IP address is valid for only a limited time, trackers can still combine it with other tracking info to track and identify internet users across multiple websites.

Cookies

Cookies have existed since the beginning of the web and are considered the most common method to track internet users. In its simplest form, a cookie is a small text file (up to 4KB) generated when a user visits a website for the first time and stored on a user computing device hard drive. Cookie content varies from one website to another and is commonly coded so that it can only be read by the website that creates it. For instance, a standard cookie file will hold information about its expiration date and the domain name that owns this cookie.

Cookies are used to perform many functions according to their types. For example, some cookies are used to maintain shopping cart contents of e-commerce sites; another type remembers the user's login, site preferences (e.g., Theme, language), and anything entered in website forms, while other types are used to record users browsing history. When a cookie has no expiration date, the web browser will delete it automatically upon close; however, some cookies have an expiration date that lasts for many years.

Cookies can be categorized according to different criteria; the most common ones are classifying them according to their expiration date (Session and Persistence cookies) and how they are used (First-party, Second-party, and Third-party cookies).

First-party Cookies

This cookie is generated by the original domain name visited by the user directly and is enabled by default in most web browsers. Website owners commonly use such cookies to collect analytics data about their visitors; they perform other valuable tasks to enhance the user experience when they return to the website (e.g., remember login info, theme, and language settings).

Second-party Cookies

This type contains all cookies that have been transferred from the First-party cookie owner (original website visited by the user) into another trusted company to use in targeted advertisements. For example, an online cloth store could sell its First-party customers' cookies to another company specializing in shoes to use it for targeted advertisements.

Third-party Cookies

Such cookies are generated by domain names other than the one visited by the user directly; their purpose is to track users browsing history across multiple websites to target them with customized Ads, among other things. This type of cookie raises serious privacy concerns, and its rejections are continually growing in the internet community.

> Note! Not all Third-party cookies constitute a privacy concern; some are used to provide services such as creating a "Live Chat" or for analytical purposes (e.g., Google Analytics).

Session Cookie

A session cookie (also known as a transient cookie) does not contain an expiration date and is stored in a temporary memory location and never stored on the user hard drive. Session cookies are erased automatically after a user closes the browser or ends the session. A good example of this type is the shopping cart functionality available in online stores. A session cookie is used to maintain the shopping cart contents and allow users to browse away from the shopping cart page while reserving the items added to it.

Persistent Cookie

A persistent cookie (also known as a permanent cookie) allows websites to remember visitors' preferences and other settings when they return in the future. For example, by using a persistent cookie, a website can remember user language preferences, theme settings in addition to saved login credentials, among other things. This increases website usability and enhances the user experience.

A persistent cookie is stored on the user computing device, and it remains there until the user deletes it from the web browser or until it expires.

A persistent cookie has mainly two types: Flash and Evercookie cookies.

- Flash Cookies (local shared object): This type is more persistent than traditional cookies with an expiration date. Flash cookies have no expiration date, and you cannot delete them by just deleting the cookies folder of your web browser, because Flash cookies are stored in a specific Adobe file on your device hard drive and can be managed via Adobe Flash Player settings. A Flash cookie can reinstall Hypertext Transfer Protocol (HTTP) cookies after a user deletes them from their web browser; it can also track users across multiple websites in addition to their ability to cross-browser tracking. Flash cookies offer the storage of 100 KB by default. However, this can be increased by the user to become infinite size; this size makes Flash cookies able to store a large amount of user data compared with the storage size of HTTP cookies.

To reveal all Flash cookies installed on your Windows device, go to Control Panel ➤ Flash Player, go to the "Storage" tab, and click the button "Local Storage Settings by Site…". To prevent storing Flash cookies on your system, select the option "Block all sites from storing information on this computer" (see Figure 2.3).

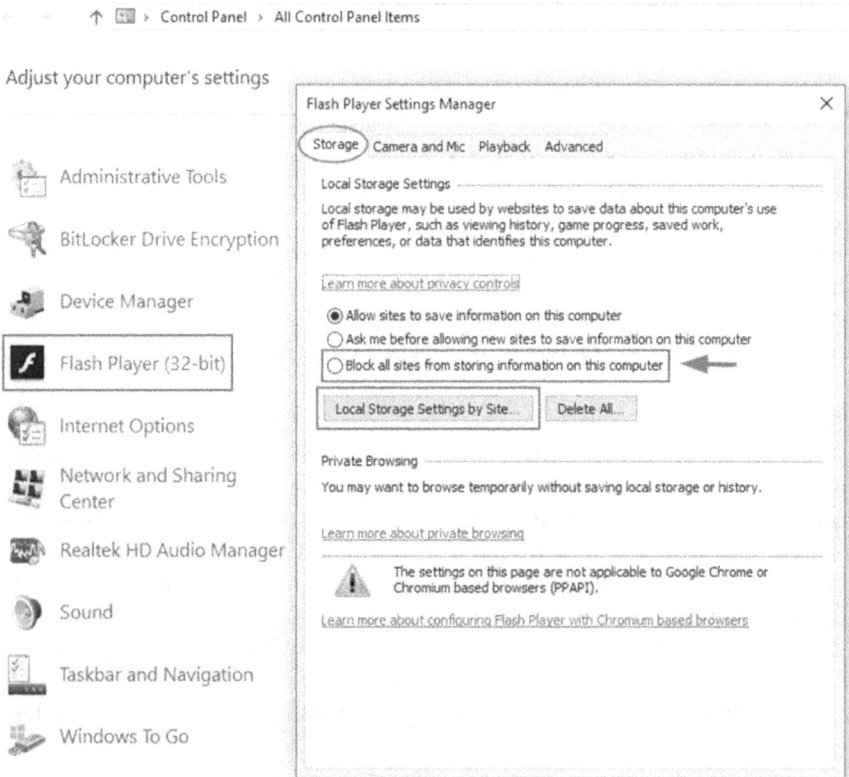

FIGURE 2.3
Checking flash cookies through Flash Player settings manager.

- Evercookie cookies: This is a highly persistent cookie developed by *Samy Kamkar* using JavaScript. Evercookie can survive even after users delete all HTTP and Flash cookies from their devices. Evercookie stores its data on different locations on user web browser/device such as Standard HTTP cookies, Flash cookies, Silverlight, Java, HTTP ETags, web cache, HTML5, and other places. If a user deletes any of these locations, Evercookie will regenerate itself again. Evercookie can also propagate between different web browsers on the same computer. To find more info about Evercookie, check its official page on GitHub https://github.com/samyk/evercookie

ETags

ETag is a shortcut for Entity Tag; this mechanism allows trackers to track online users without using cookies, IP address, JavaScript, Java, or any LocalStorage/SessionStorage/GlobalStorage object. To learn how ETag can be utilized to track internet users, we should first understand how the web works.

When a user (client computer) wants to visit a webpage or download a file from the internet, it requests this resource from the web server using the HTTP language. After receiving the request, the web server will respond to the client and send the requested resource. To speed up the process and make it more efficient, new mechanisms were developed; the most important one was Web Cache.

ETags is part of an HTTP mechanism that provides web cache validation. It compares local resources stored on the user device with those stored on the web server to determine whether it needs to download these resources when requested again. This makes the connection faster and saves bandwidth.

The bad side with ETags is their ability to track online users across different sessions even though the user has deleted browser cookies and disabled all JavaScript and Flash functionality. This can be achieved technically when the tracking server continually sends ETags to a client browser, even though the contents do not change on the server. By doing this, a tracking server can maintain a session with the client machine that persists indefinitely (see Figure 2.4).

The best solution to stop ETag tracking is to delete the web browser cache on exit, although this will not eliminate ETag tracking completely.

Device Fingerprinting

Also known as "Browser Fingerprinting", in this type, individual computing devices are identified based on their browser and hardware/software configurations. This tracking technique is stateless and transparent to the user and device. Web browsers reveal much technical information about

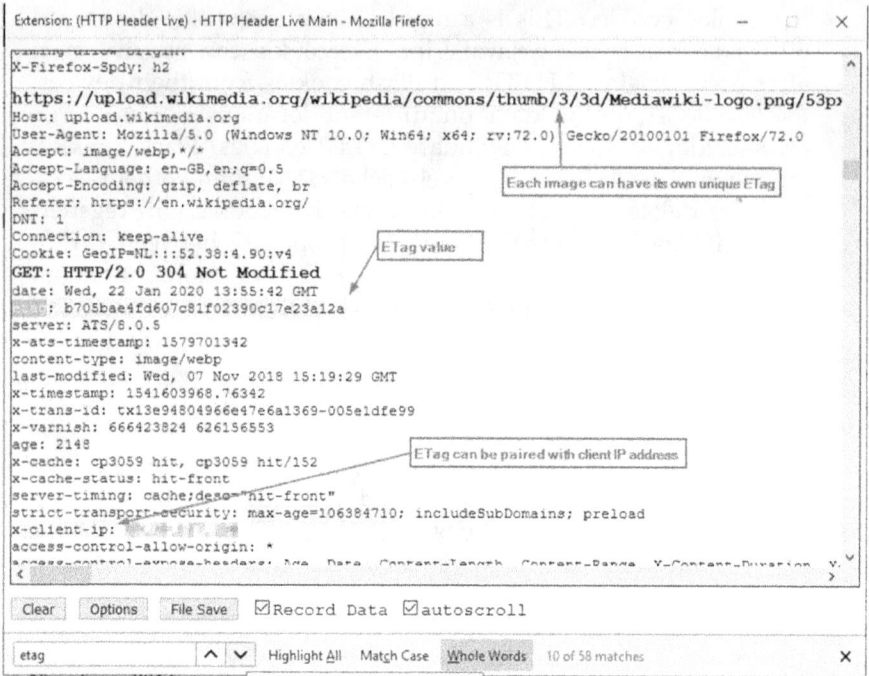

FIGURE 2.4
Viewing Wikipedia header info using Firefox browser extension HTTP header live (https://addons.Mozilla.org/en-US/firefox/addon/http-header-live).

your computing device settings to help websites understand how to display their contents based on browser version, operating system, and screen size. This technical information is what makes up your unique digital fingerprint when surfing the internet.

Trackers collect different technical information about users' devices, such as:

- Screen resolution,
- Operating system type,
- Browser type and version,
- Time zone,
- Location,
- Installed fonts,
- Cursor scrolling behavior,
- Installed web browser add-ons,
- Battery information,
- Device language.

The collected technical parameters are then used to create a unique digital fingerprint of the user device. This fingerprint facilitates profiling internet users by tracking all their online activities. You may think such technical information is generic and cannot identify individual devices among millions of connected devices; however, this is not true. If we combine this info, we will have a unique fingerprint of the user's device. This device fingerprint can be later associated with real user identity if combined with other Personally Identifiable Information (PII).

Note! What is PII?
PII is any information that can be used on its own to identify the person. Examples of PII include: Full name, email address, login ID, phone number, Social Security number, bank account number, driver's license number, passport number, biometric records, or any other information that uniquely belongs to you and is personally identifiable.

Digital fingerprinting allows trackers to profile internet users without leveraging traditional tracking mechanisms such as IP addresses and cookies. The Electronic Frontier Foundation (EFF)[3] conducts an experiment on web browser fingerprintability of half a million distinct browsers. The result concludes that most internet users could be uniquely fingerprinted and tracked using only their web browsers' configuration and version information.

Although EFF used a limited set of parameters to fingerprint users' devices, it found that 84% of the devices had unique configurations. In comparison, the percentage rise to 94% when target devices have Flash or Java installed. Commercial fingerprinting services use more advanced measurements to fingerprint devices. Obviously, this will increase the percentage of devices that can be precisely identified online.

Device fingerprinting can be technically implemented using two methods: Script and Canvas fingerprinting.

Script Fingerprinting

As its name implies, a script is loaded into the user's web browser and then executed to extract various technical parameters of the current browser and device settings. A hash is then made based on the information collected from the script and is used to track user devices across the internet. The script used to fingerprint devices is usually developed using JavaScript programming language. However, other languages can perform the same thing, such as Flash, Silverlight, or a Java applet.

Canvas Fingerprinting

Hypertext Markup Language (HTML) is the standard markup language used to build websites and web applications accessed on the internet. HTML5 is the last major version of HTML. Canvas is an HTML5 coding element initially developed by Apple and is used to draw graphics and animation on web pages via JavaScript API. Canvas can be exploited to fingerprint web browsers and thus, track users' online activities.

Canvas fingerprint devices by drawing a text into the canvas of the user's browser and then reading the rendered – invisible – image data back. Even though the same image was used on different client devices, it will be rendered differently because each will have a different set of software and hardware configurations. Once finished, this image will play the role of script fingerprinting in obtaining various technical information about the user device hardware and software stack. Finally, a hash is made based on the generated Canvas data and used to distinguish a user client device across the web.

Web Beacon

A web beacon (also known as a web bug, pixel tag) is an invisible (transparent) and tiny (pixel in size) graphic image embedded in a website or included within an email message. Although beacon is mainly used for analytical purposes by website owners, advertising companies utilize it to track online user activities to deliver targeted advertisements. Beacon is widely used to track HTML email messages. It works by placing a 1x1 pixel image within the email; when the recipient opens the tracked email, the beacon will get activated and extract various technical info about the recipient computing device. Such as the date/time when the email was opened, when a user clicks on a link within the email, IP address of the device – or the IP address of the VPN/Proxy service – used to read that email, type of browser/email client used to read the email, and OS type. This info is then transferred and recorded in a log file by the tracking server.

A Web beacon is usually used in combination with cookies to facilitate third-party tracking services. It can also retrieve the user IP address among the collected technical info making it able to track user browsing activities similar to cookies.

Although beacons are famous for using graphic images for tracking, beacon technology has developed to include more elements such as banners, buttons, and non-pictorial HTML elements such as frames, scripts, and input links.

Note! Spammers commonly use a web Beacon to validate email addresses. When a user opens an email message that contains a web beacon, the spammer gets notified that this email address is valid.

Examples of Online Tracking

There are many examples of online tracking; having said that, the most threatening for internet privacy is third-party tracking; in this section, we will talk about two popular – and widely used – third-party tracking examples.

Social Networking Tracking

This is the most popular form of online tracking; it is usually conducted by giant IT companies and social media platforms to profile internet users on a mass scale. Facebook and Twitter are clear examples.

Facebook can track users internet browsing activity, whether they are logged to its platform or not! Facebook achieves this technically via the "Like" and "Share" buttons already spread everywhere around the internet. Facebook inserts its tracking – JavaScript – code inside these controls; when a user visits a website containing these buttons, Facebook will automatically track and record user activity even though they are not signed into Facebook.

> Note! Facebook and Twitter social buttons place cookies on users' devices, and these cookies are third-party cookies because they can track users across multiple websites.

For websites that do not contain Facebook "Like" & "Share" buttons, Facebook can still track users' internet activities by using a tracking mechanism known as Facebook Pixel (www.facebook.com/business/help/ 742478679120153). Facebook Pixel is a JavaScript code installed on websites – similar to the Web Beacon concept; it is a kind of analytics tool, however, Facebook can use it to track and record users' internet browsing history for targeted advertising.

> Note! All Facebook companies such as WhatsApp, Oculus, and Instagram share data collected by Pixels. This gives Facebook broad access to most internet users browsing history.

Twitter "Follow" button can also track online users just like Facebook "Like" and "Share" buttons do.

Search Engine Tracking

Popular search engines such as Google, Yahoo!, and Bing are known to store users' search queries and track them – even when they are using a private browsing mode – across the internet to profile and target them with custom-ized ads.

For example, most Google search engine users already own a Gmail account (Google free email service). When any user uses Google search while logging into Gmail, Google will be able to record users browsing his-tory and link it to their real identity (as most Gmail users already use their real identity – or a valid phone number – when signing up for a Gmail account). Even though a user has not logged into their Gmail account, Google can still track their online activities and link it to user's real identity using any of the tracking techniques already mentioned (e.g., cookies, digital fingerprinting).

> Note! According to Princeton's WebTAP privacy project (https://www.cs.princeton.edu/~arvindn/publications/OpenWPM_1_million_site_tracking_measurement.pdf), the most three websites that have the larg-est tracker networks installed on the majority of internet websites are:
>
> 1. Google, it has trackers installed on 75% of the top million Internet websites
> 2. Facebook 25%
> 3. Twitter 10%

To stay safe and assure your privacy online, you can use a search engine with privacy-enhanced features. The following list some popular search engines which do not record online user activities:

- DuckDuckGo (https://duckduckgo.com)
- Startpage (https://www.startpage.com)
- Disconnect Search (https://search.disconnect.me)
- Qwant (https://www.qwant.com)
- Lukol (https://www.lukol.com)
- MetaGer (https://metager.org)

Protection Methods

As we saw, tracking internet users is becoming widespread, there are differ-ent tracking mechanisms currently employed, and possible new means will continue to emerge in the future.

Evading online tracking is hard to achieve and requires a user to have good technical skills if the purpose is to achieve 100% anonymity, nevertheless, for most users, achieving complete anonymity online is not their ultimate goal.

They simply need to ensure their online privacy and prevent others from recording their browsing history and linking it to their real identity.

In general, to prevent online tracking, you should implement the following three main countermeasures:

1. Conceal the real IP address used to access the internet.
2. Delete cookies and browser cache upon closing the browser.
3. Prevent device fingerprinting techniques from profiling your computing device and consequently tracking your internet activities.

We will expand each countermeasure in this section.

Disable Third-Party Cookies

Deleting and blocking cookies is the main countermeasure against tracking cookies; however, keep in mind that 1st party cookies do not impose a privacy risk compared with 3rd party cookies installed by domains other than the one visited directly by the user.

To prevent trackers from tracking you using third-party cookies, you can activate the privacy mode in your web browser.

Use Incognito Mode

To enable private browsing in Firefox, open Firefox and click **CTRL + SHIFT + P** keyboard shortcut, a new browser window appears stating you are in private browsing mode. You can also right-click on any link and choose "Open Link in New Private Window" from the context menu. When surfing in Privacy mode, Firefox will not store your browsing history and delete all cookies at the end of your private session.

Google Chrome offers similar functionality to browse in private called "Incognito mode". To activate this mode:

1. Open Chrome.
2. At the top right, click More, and then "New Incognito Window" (see Figure 2.5).

Alternatively, you can activate this mode under Windows, Linux, or Chrome OS by pressing **Ctrl + Shift + n** keyboard shortcut.

Examine Third-party Cookies Via Web Page Source Code

We can also check if a particular website uses Third-party cookies by following these steps (I'm using Google Chrome browser):

1. Open Developer Tools by pressing F12.
2. Choose the "Application" tab.
3. Double-click the "Cookies" section on the left pane to expand it.

FIGURE 2.5
Activate incognito mode in Google Chrome.

4. The current domain name and also subdomain name cookies should appear here.

Any cookies belonging to a domain name other than the one you are visiting are third-party cookies (see Figure 2.6).

Disable Third-party Cookies Under Firefox

Firefox has an option to disable third-party cookies. We can enable it as follows (I'm using Firefox version 95):

1. Go to the *Tools* menu ≫ *Settings*
2. Select the *Privacy & Security* panel.
3. Under *Enhanced Tracking Protection*, select the *Custom* radio button to choose what to block.
4. To disable all third-party cookies, select the *Cookies* checkbox and select *All third-party cookies (may cause websites to break)* from the dropdown next to it (see Figure 2.7).
5. Close the Settings page to save all changes automatically.

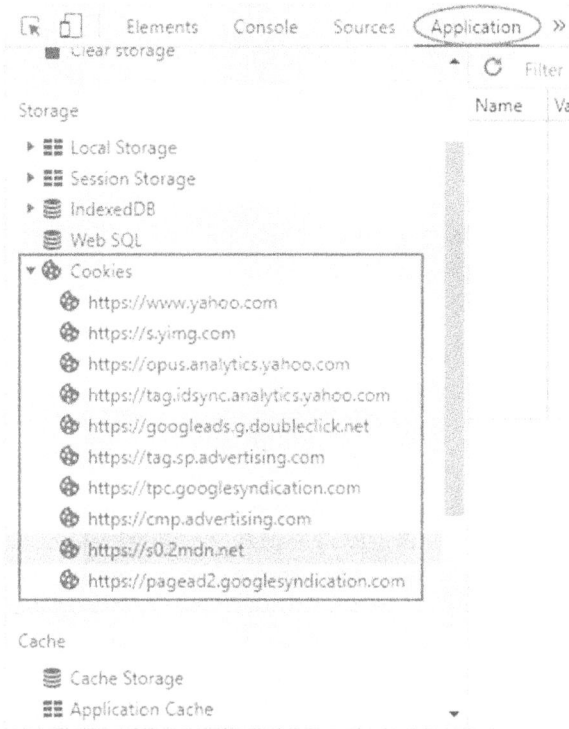

FIGURE 2.6
Viewing third-party cookies using the developer tools under Google Chrome for the Yahoo. com domain name.

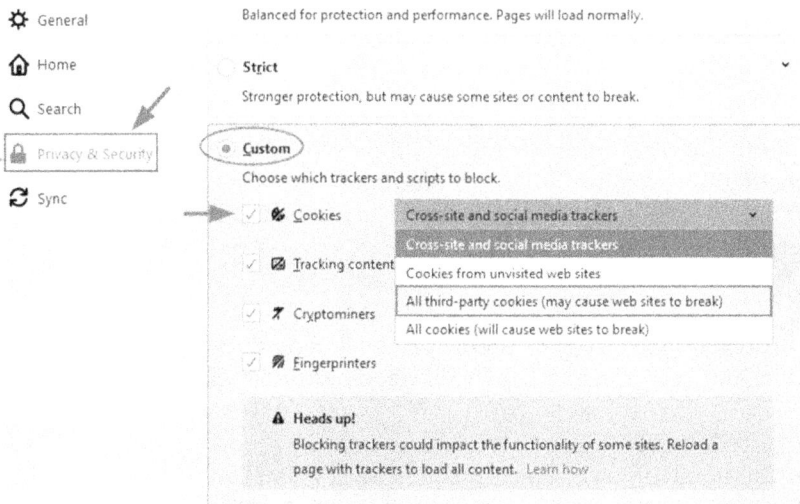

FIGURE 2.7
Disable third-party cookies under Firefox.

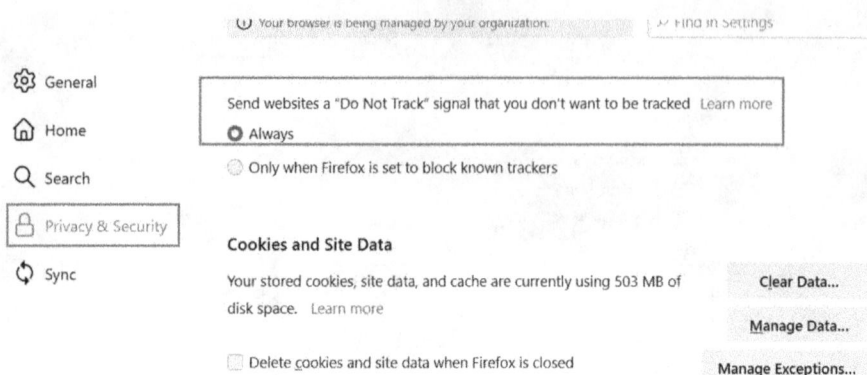

ⓘ Your browser is being managed by your organization.   ↗ Find in Settings

⚙ General

🏠 Home

🔍 Search

🔒 Privacy & Security

🔄 Sync

Send websites a "Do Not Track" signal that you don't want to be tracked Learn more

◉ Always

◯ Only when Firefox is set to block known trackers

Cookies and Site Data

Your stored cookies, site data, and cache are currently using 503 MB of disk space. Learn more

☐ Delete cookies and site data when Firefox is closed

Clear Data...

Manage Data...

Manage Exceptions...

FIGURE 2.8
Activate the "Do not track" option.

From the same section, we can also activate the option "Send websites a 'Do Not Track' signal that you don't want to be tracked" and set it to "Always" (see Figure 2.8). You can learn more about this feature by going to https://support.mozilla.org/en-US/kb/how-do-i-turn-do-not-track-feature

Test Your Browser's Fingerprinting

It is helpful to see what your current OS/browser configuration reveals about your device. There are many free online services to check the digital fingerprint of your device and test if it is safe against tracking.

1. PANOPTICLICK (https://panopticlick.eff.org)
2. AmIUnique (https://amiunique.org)
3. Browserleaks (https://browserleaks.com)

Countering Browser Fingerprinting

Countering browser fingerprinting is the most difficult among other tracking methods, as you need to incorporate all mitigation techniques against all other online tracking methods and more to prevent it. To fight against device fingerprinting, follow these steps:

Surf Using Incognito Mode

We already covered how to enable privacy mode in both Firefox & Google Chrome and how to reject Third-party cookies under Firefox.

Install Browser Plugins to Stop Trackers

There are many browser extensions for privacy that can disable online trackers from running scripts in your browser and do their work in identifying your online footprint harder. Remember that some extensions may make your browsing experience less satisfactory as some websites' features may get broken; however, you can avoid this by disabling add-ons on the websites you trust.

1. Privacy Badger – block invisible trackers, works on major web browsers (https://www.eff.org/privacybadger)
2. uBlock Origin (Firefox and Chrome) – efficient tracker blocker for Firefox (https://addons.mozilla.org/en-US/firefox/addon/ublock-origin)
3. Adblock Plus – open-source add-on, blocks intrusive ads and stops different online trackers for both Firefox and Chrome (https://adblockplus.org)
4. Disconnect – Visualize and block invisible websites that track your search and browsing history for Firefox (https://addons.mozilla.org/en-US/firefox/addon/disconnect)
5. Cookie AutoDelete for Firefox – Auto Deletes Cookies upon closing each tab (https://addons.mozilla.org/en-US/firefox/addon/cookie-autodelete)
6. NoScript (both Chrome and Firefox) – allows JavaScript, Java, Flash, and other plugins to be executed only by trusted websites of your choice (https://addons.mozilla.org/en-US/firefox/addon/noscript).

Disable JavaScript and Java

The most common device fingerprinting technique works by executing JavaScript code on the user's browser to retrieve technical information associated with it. Disabling JavaScript will break the functionality of many websites as most web frameworks are now built using JavaScript. To solve this problem, a user can install the NoScript browser add-on and whitelist every trusted website and disallow all other websites from running the following scripts: JavaScript, Java, Flash.

Some fingerprinting techniques use Java from Oracle to fingerprint user devices; we can disable Java in the web browser by doing the following under Windows OS:

1. Go to Control Panel ≫ Java
2. Click on the "Security" tab
3. Uncheck the checkbox for "Enable Java content in the browser" (see Figure 2.9). This will disable the Java plugin in the browser.
4. Click "Apply", then "OK".

Countermeasures Against Web Beacon Tracking

Web beacons cannot be avoided easily; you can discover if a website is using beacons by checking the web page source code. If you find any image tags that load from a different domain than the rest of the site, this could be a beacon tracking image. Rejecting cookies will also prevent beacons from tracking your activity.

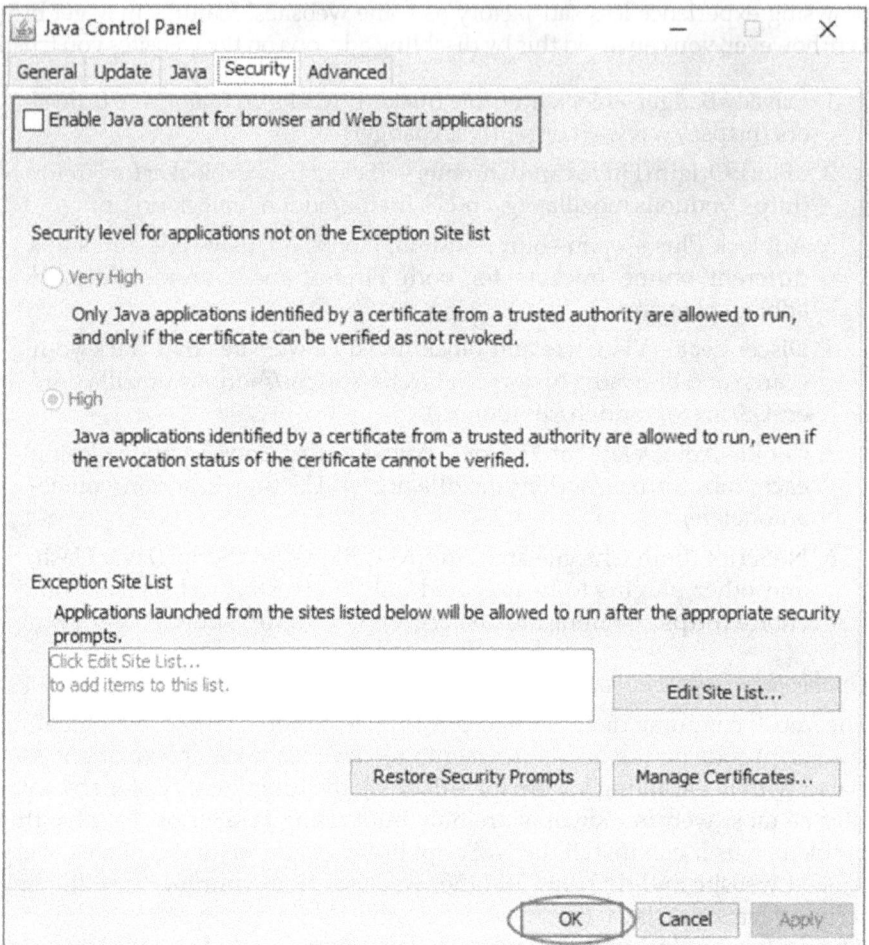

FIGURE 2.9
Disable Java for web browsers.

Email beacon (delivered via HTML email) can record the same technical information about the client device, similar to the beacons installed on websites. To stop email tracking by beacons, do the following:

1. Disable viewing remote contents in an email client. For instance, most email clients like Outlook and Mozilla Thunderbird can be configured to avoid opening remote images in received emails. By default, Thunderbird does not load remote content; however, to make sure this is the current configuration in your Thunderbird client (I'm using version 91), go to the *Tools* menu ≫ *Preferences* ≫ Privacy & Security, make sure the option *"Allow remote contents in messages"* is unchecked (see Figure 2.10).

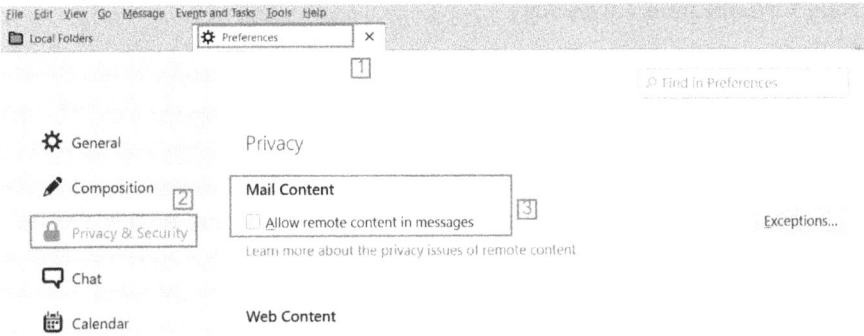

FIGURE 2.10

Disable remote contents viewing in emails under Thunderbird email client (v68.3) – Cookies and web contents can be configured from the same window.

2. Another option to avoid email beacons is to download the email messages first using your preferred email client (e.g., Outlook or Thunderbird) after then disconnect your device from the internet before you begin reading the messages.

3. The last and most secure option is to display all received emails in Plain Text only. A web beacon cannot execute and communicate with its host server when the HTML formatting is disabled in the recipient email client. To configure Thunderbird to change its default email display to Plain Text: go to the *View* menu ≫ *Message Body As* ≫ *Plain Text* (see Figure 2.11).

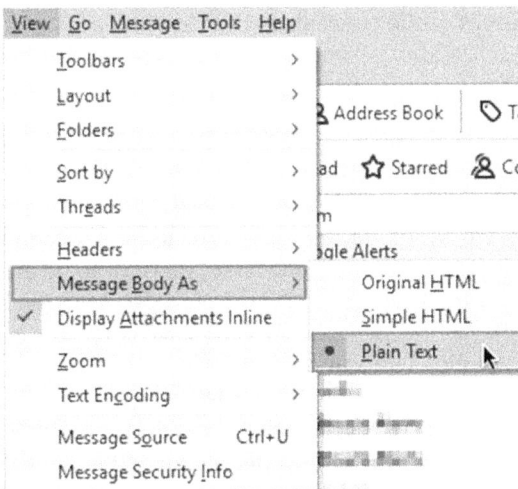

FIGURE 2.11

Configure Thunderbird to display messages in plain text.

Using Virtualization Technology

We have already listed different techniques to prevent trackers from identifying internet users and recording their browsing history; however, all the precautions already mentioned cannot guarantee 100% of privacy from trackers when going online. For instance, non-Internet-savvy users can still make some mistakes while surfing, leading trackers to recognize their devices.

The best technical solution is to make your device's digital fingerprint similar to most devices' fingerprints! To achieve this, you need to use a freshly installed OS and web browser configured with solid privacy settings without installing any extensions. You also need to use a reliable VPN service to mask your actual IP address. This OS should be installed on virtual machines like VirtualBox https://www.virtualbox.org.

By doing this, your device will not become unique when going online, and your OS/Browser fingerprint will look similar to millions of other connected devices.

Note! There are many guides to configure Firefox settings to become more privacy-friendly. **Restore privacy** has an excellent one: https://restoreprivacy.com/firefox-privacy/

Now that we know how web tracking works and the best method to prevent it, we can start to learn how to become anonymous online to prevent outside observers from attributing our online actions to our real identity. This knowledge is a prerequisite for security professionals wishing to access the darknet or gather intelligence about criminal or terrorist activities from online public sources while concealing their real identity.

Online Anonymity

Like the physical world where people's right to be anonymous is protected as part of free expression laws, people have the right to be anonymous in the digital world. However, this right is hard to achieve because of the boundless nature of the digital world, where different actors are interested in unmasking users online activities for various reasons.

Online anonymity is a complex topic that triggers a long debate between security researchers, corporations, politicians, and the public. This debate is summarized in one question: Should people have the right to anonymity on the internet? Or not?

The internet was not designed at first to be an anonymous network, but this was not an issue in its early days. However, as more people are entering the digital world and begin to use the internet in their daily interactions, the issue of anonymity on the internet becomes increasingly essential for public users. Unfortunately, like most things in life, bad actors have exploited the ability to be anonymous online to conduct criminal actions that they cannot do in person.

As a cybersecurity professional, it is essential to understand how online anonymity can be achieved to protect your identity online and discover how cybercriminals can exploit anonymity services to conceal their true identity when commenting crimes. For instance, different anonymity services were developed to ensure end-users' privacy online. This section will list the two main methods to conceal your identity online: Anonymous networks and VPN.

Anonymous Networks

An anonymity network is a closed network of anonymous computers spread all over the world that form a decentralized network, such networks can only be accessed via specialized software or by configuring your web browser to use specific ports/protocols which is different from the one used to surf the ordinary internet. Some anonymous networks – such as the TOR network – allow their users to access and surf the regular web (also known as the surface web) anonymously through encryption.

There are many anonymous networks; however, only a few of them are popular and reliable for daily internet surfing, as we will see in the "Dark web" section.

VPN

VPN, stands for Virtual Private Network, allows you to establish a secure tunnel from your computing device to another destination online. Everything you exchange with the destination (e.g., a remote server) is encrypted and cannot be intercepted by outside parties (such as your government or ISP). In addition to encrypting internet connections, VPN allows internet users to conceal their IP addresses and access websites censored by their ISPs.

VPN was initially developed to securely facilitate connecting organization networks scattered in different geographical areas over the internet and provide a secure channel for employees wishing to connect to their corporate network from home. However, these days, VPN usage has become very popular among internet users to encrypt internet connections and prevent eavesdropping, especially when using public Wi-Fi hotspots.

VPN works by connecting your computing device (laptop, smartphone, tablet, desktop) to the VPN server located online (usually in another country other than the user's country). The VPN server will conceal your actual IP address and give you another IP address to surf the web privately without exposing your private information and browsing habits to outside observers.

Please note that not all VPN service providers are equal in terms of security and reliability. For instance, a VPN provider can intercept everything that passes through its channel, such as login credentials, visited websites, and the actual IP address of the user. For these reasons, you should choose your VPN provider very carefully and avoid using any free VPN service. To help you better select the best VPN provider, the book author advises you to check his detailed guide about VPN selection criteria available at https://www.secjuice.com/how-to-choose-a-virtual-private-network-vpn-provider

---

Define Web Layers

The general public thinks the web is what they see when using Facebook, watching YouTube movies, checking email, or searching using Google or other similar search engines. These places are considered a part of the surface web where most internet users' daily online activities take place; however, what the general public is accessing barely constitutes 4% of the whole World Wide Web!

Web can be categorized into three layers concerning content accessibility: Surface web, deep web, and the dark web. The dark web layer is a sub-layer of the deep web; however, it is usually included in a separate group because of the technical complexity associated when accessing and surfing it.

Note! The general public refers to the World Wide Web as the Internet; however, this is wrong. The web is composed of interlinked HTML pages accessible via the internet connection.

Surface Web

Also known as the Clearnet, this is what most people see when going online. The surface web contains all web contents that can be indexed and discovered by conventional search engines such as Bing, Yahoo!, and Google. Despite being accessible by the general public easily, the surface web occupies less than 4% of the entire web content.

What makes the surface web usable by most people is search engines. Next, we will see how search engines index web content.

A search engine sends a crawler – also known as spider or robot – to discover updated and new contents online, this spider visits each domain name and crawl its contents – which include most popular file types such as image, video, PDF and Office documents – by following hyperlinks between webpages, it then sends the discovered content to the search engine server to

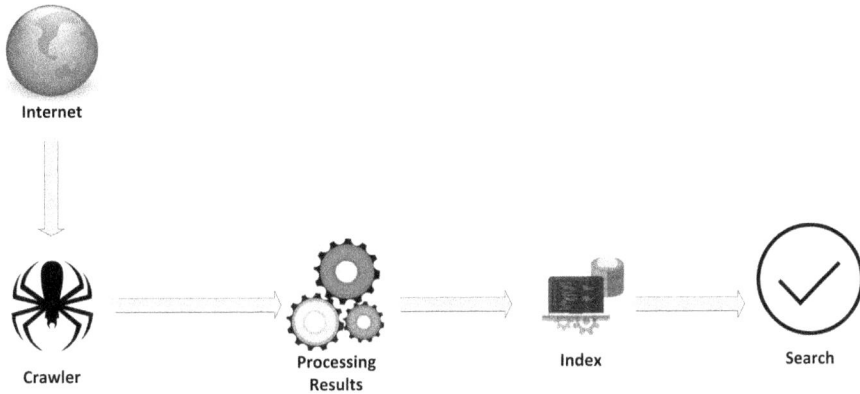

FIGURE 2.12
How search engine works.

process it and organize the results obtained from crawling process into an index. The index is a massive database file containing all discovered results that can exceed trillions of web pages. Finally, when a user submits a search query, the search engine will return the best results to answer user queries and order them appropriately, beginning with the most relevant and ending with the least relevant (see Figure 2.12).

Deep Web

As mentioned earlier, search engines discover new content by clicking hyperlinks; however, this technique is not perfect as there are places on the web that do not have a hyperlink point to, hence, cannot be discovered – or accessed by web crawlers.

Standard search engines are unable to index the deep web for the following reasons:

1. A large amount of deep web data is protected with a password (restricted content). It needs some sort of authentication to view it. For example, the Taylor & Francis website https://www.tandfonline. com holds millions of articles and scientific papers, many of which require a user to have an account to access. This digital library is a clear example of content that web crawlers cannot access.

2. Dynamic pages require a user to submit a search form or use the drop-down menu to set some search values to appear. For example, let us assume we want to see whether the historical information of New York City on February 9, 2011. Many websites offer

historical weather data. For this example, I will go to https://www.wunderground.com/history and enter the name of the city I want to search. I also need to adjust the date in the search form to point to my desired date, which is 2011/02/09 for this example. Finally, I need to click the "View" button to submit my search query and retrieve the results from the database. If I act as a search engine crawler, I would only click on links on the page without filling any search query and submitting search forms; obviously, this will not return any result.

3. Some website owners choose not to index their webpages by placing a robot.txt file in the root directory of their website; this instructs web crawlers not to index the site, making it a part of the deep web. On the other hand, there are also some file formats that search engines do not index. For example, web crawlers cannot index text within graphics file formats (e.g., JPEG, GIF, or TIFF) and archived contents with customized extensions.

4. The fourth factor that effect on search engines ability to index the deep web is that the data resides on encrypted networks, such as the websites hosted on the darknet networks (covered next) such as TOR and Invisible Internet Project (I2P) because it cannot be accessed by typical search engine crawlers.

Note! A list of Google Indexable File Formats can be found at https://support.google.com/webmasters/answer/35287

The deep web constitutes the largest portion of the web, with nearly 95% of the contents residing on this layer. Although it is less accessible to the general public, deep web contents are of exceptional value for corporations, educational institutes, and government organizations because their high quality and authoritative resources make it extremely valuable for focused searches.

Some popular deep web websites:

1. The Library of Congress (https://www.loc.gov). This is the most extensive online library globally with topics covering all life aspects such as books, journals, films, newspapers, maps, software, etc.

2. Science.gov (https://www.science.gov). This website holds over 60 databases and over 2,200 scientific websites that contain more than 200 million pages of authoritative federal science information, including research and development results.

3. Google Books (https://books.google.com). A database that holds millions of digitized textbooks covering all science subjects.

4. Wayback Machine (https://archive.org). Gives historical information about any URL. It also archives millions of books, movies, and software uploaded by users.

Searching the deep web manually is a daunting task for individuals; however, many online services facilitate searching deep web content for niche subjects. Here are some:

1. Google Scholar (https://scholar.google.com). Facilitates searching the deep web for scholarly literature.
2. TruthFinder (https://www.truthfinder.com). A people search engine that searches social media, photos, police records, background checks, civil judgments, contact information, and more. Its results are extracted from different deep web databases.
3. Shodan (https://www.shodan.io). Shodan is the world's first search engine for Internet-connected devices.
4. Blogspot Blog Search (https://www.searchblogspot.com). A blog search engine to search for blogs created on the Blogger platform.
5. Marine Traffic (https://www.marinetraffic.com). Track vessel movements in real time. It also provides historical information about any vessel worldwide.

Dark Web

Also known as the Dark Net, this is the deepest layer and is considered a part of the greater "deep web" portion. Unlike the deep web that can be accessed using a regular web browser without needing any particular program or configuration, the dark web requires special software (such as the TOR browser) to access. The dark web is not one network; it is a term used to name hidden networks which are composed of computers running by volunteers from different places around the world that work together to form a decentralized web. The collection of anonymity networks forms what is known as the Dark web.

Dark web is built for anonymity, and the websites hosted within it are made purposefully hidden. This portion of the internet is famous for being associated with illegal activities.

Note! People use the term Deep Web and Dark Web interchangeably, however, this is not accurate. The dark web is a tiny subset of the deep web that no one can predict its size because of the ephemeral nature of many of its websites.

The strong anonymity offered by the darknet attracts criminals to use it as a medium to sell/buy all sorts of illegal products and services. Digital black markets on the darknet sell arms, drugs, stolen credentials, leaked financial secrets, child pornography, copyright materials, hacking services, false government documents, and anything forbidden in the real world. Traders on the darknet use cryptocurrencies – such as Bitcoin – to keep their payment transactions anonymous.

Although a large portion of the darknet websites is involved in illegal activities, a small part of them is utilized for noble purposes. For instance, human rights activists and journalists living in countries controlled by oppressive regimes use the darknet to promote their ideas and exchange information with the outside world without the fear of being tracked by authorities. The darknet is not merely used to host anonymous websites. For instance, internet users seeking privacy when browsing the Clearnet (surface web) can use the TOR anonymity software to anonymize their internet connection and prevent trackers from tracking their online activities. Examples of such user groups are whistleblowers leaking info and individuals wishing to circumvent government censorship on forbidden websites.

Popular Darknet Networks

As we already mentioned, the Dark web comprises a collection of anonymous networks of varying size and popularity; some of these networks are managed by individuals, while public organizations sponsor others. The most popular anonymity networks are TOR and I2P.

TOR

TOR is an anonymizing network that directs internet traffic over a series of volunteer computers spread around the internet. TOR is famous for being used as a proxy to conceal the internet user's identity (IP address) when surfing the surface web.

Initially developed by the United States Naval Research Laboratory in the mid-1990s with the purpose to protect online intelligence communications, TOR has undergone many development processes; the most notable one was in 2004 when the Naval Research Laboratory released TOR source code, making it available under a free license. TOR is currently managed by The Tor Project, Inc., which is a nonprofit organization responsible for TOR development.

How TOR Works?   In simple terms, TOR works by routing your internet connection from your computer to the destination website (e.g., www.OSINT. link) using a set of intermediate servers or relays (mostly three relays), the connection between these relays is encrypted.

TOR relays run by volunteers worldwide who offer a part of their connection bandwidth for the TOR network. To participate as a relay, you need to have the TOR software installed on your device and configure it to act as a relay.

TOR Relay Types

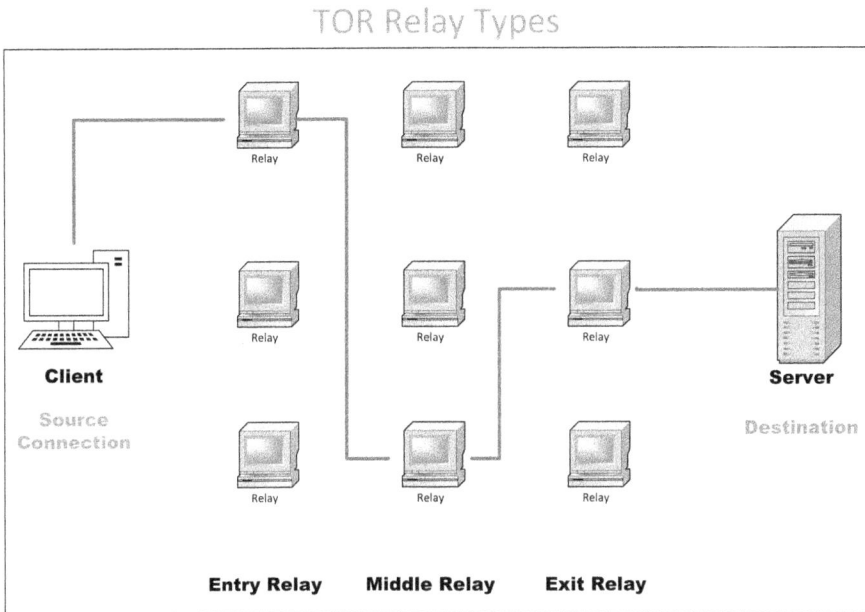

Client

Source
Connection

Server

Destination

Entry Relay   Middle Relay   Exit Relay

FIGURE 2.13
How TOR tunnel traffic.

We can distinguish between three types of TOR relays (see Figure 2.13):

Entry relay: Also known as the guard node, this is the entry point to the TOR network. At this point, a user accesses the TOR network coming from the surface web. It is advisable to use a VPN connection – or TOR bridge/pluggable transports – to hide your entry to the TOR network from your ISP.

Note! To conceal your entrance to the TOR network and if TOR itself is blocked in your country, use the TOR bridge or pluggable transport to overcome this obstacle. TOR provides a detailed guide on using this on the following URLs: https://blog.torproject.org/tor-heart-bridges-and-pluggable-transports & https://tb-manual.torproject.org/circumvention

Middle relay: As its name implies, the middle relay stands in the middle of the TOR connection, it is responsible for moving traffic from the Entry relay to the Exit relay. This effectively hides the original source of the connection from the Exit relay.

Exit relay: This is the last node in the TOR network; it sends the traffic out of the TOR network to its final destination. Exit relays can intercept traffic as the data needs to be decrypted to reach its final destination on the regular internet. To mitigate this risk and prevent malicious TOR Exit relays from eavesdropping on your traffic, it is essential to encrypt your data and use HTTPS when connecting to websites on the surface web using the TOR connection. This is why the TOR browser comes with the HTTPS Everywhere add-on (https://www.eff.org/https-everywhere) preinstalled.

Websites hosted on the TOR network are known as TOR hidden services; they have. **onion** extension and cannot be accessed from the surface using regular web browsers. TOR websites use TOR technology to remain secure and anonymous unless the owner wishes to make them public.

The speed of the TOR connection is relatively slow compared with regular internet connection speed; this is because TOR depends on its volunteer computer devices for bandwidth, which is limited. As a result, the more relays in use, the better the bandwidth and speed of the TOR connection.

TOR is the most important and popular dark web anonymity network and is used widely for anonymizing internet connections when accessing the surface web, so it will be our focus in this chapter.

I2P

This is the second popular anonymity network after TOR. I2P stands for the Invisible Internet Project; first released in 2003, I2P is free and open-source software written using the Java programming language. Unlike TOR which is designed to provide strong anonymity when surfing the surface net, I2P was designed to focus on securing internal communication. I2P is a decentralized anonymous peer-to-peer network above the surface web where users can exchange communications within the I2P network securely and anonymously using four layers of encryption (garlic encryption[4]).

I2P runs on tens of thousands of volunteer computers around the world, each I2P node is named a router, and each router provides a one-way encrypted connection between other users within the I2P network.

I2P websites are known as *eepsites*, it has a **.i2P** extension and can only be accessed via the I2P network. After running the I2P software on your machine, you can configure various applications to tunnel their traffic via the I2P network, including torrent, email, internet messaging, and web browser applications.

Both TOR and I2P are anonymous networks; however, they differ in their design goals. For instance, I2P design goals were to create an anonymous network that provides complete anonymity for users within the I2P network; I2P was not invented to be used as a proxy to anonymize access to the surface web. For instance, I2P has a low number of outproxies (equivalent to TOR Exit relay) to the regular internet, making it ineffective – compared to TOR – in concealing the user's identity when leaving the I2P network to the internet.

Accessing TOR Darknet Network

To access the TOR network, all you need to do is to download the TOR browser, a portable security-hardened Firefox browser that comes bundled with the TOR software. Using this browser will automatically tunnel all your internet traffic through the TOR network. This browser has versions for all major operating systems; you can get your desired version from: https://www.torproject.org/download

However, before you access the TOR darknet (this also applies to any other darknet), a set of precautionary steps should be followed strictly to avoid revealing your true identity.

How to Safely Navigate the Darknet?

1. Make sure you download the TOR Browser from the official website only.
2. Hide your entry to the TOR network using a reliable VPN service.
3. Make sure you are using the latest version of the TOR browser.
4. Ensure your OS and the installed applications – including your security software – are all current.
5. Ensure to disable JavaScript in your TOR browser, and do not install any add-on except the one already installed by the TOR browser (TOR browser comes with two add-ons already preinstalled by default: HTTPS Everywhere AND NoScript).
6. For mission-critical tasks, it is advisable to change your TOR current IP address for each website visited. You can do this by clicking on the hamburger icon in the top right and selecting "New Identity" (see Figure 2.14). Take note, this will restart TOR browser and disconnect any opened sessions.
7. Use a specialized email address for the TOR network. Do not use your regular email address for registration on TOR websites, and do not access any private account – linked to your real identity – via the

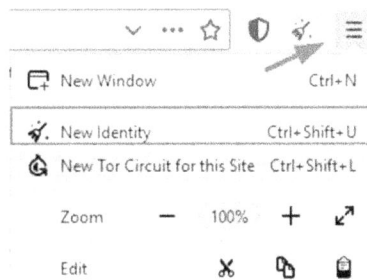

FIGURE 2.14
Changing TOR browser identity.

TOR browser. ProtonMail utilizes TOR at the following onion address: https://protonmailrmez3lotccipshtkleegetolb73fuirgj7r4o4vfu7ozyd. onion

8. Do not use your real name or any piece of personal information that points to your real identity when registering on darknet websites.

9. Do not open PDF files using your TOR web browser; instead, download them to your computer and open them using your regular PDF reader.

10. Cover your computer Camera and Microphone.

Searching TOR Network

Finding information on the TOR darknet is more challenging than using Google on the surface web, this is because search engines cannot index darknet contents. There are many dark web search engines; however, dark-net websites' hidden – and ephemeral nature makes indexing their contents almost impossible to achieve. Nevertheless, some TOR websites try to aid with TOR searching; the following are the most popular:

1. Ahmia (http://juhanurmihxlp77nkq76byazcldy2hlmovfu2epvl5ank dibsot4csyd.onion). You can access *Ahmia* from the surface web using the following address: https://ahmia.fi; however, you still need to use the TOR browser to access **.onion** websites.

2. The Uncensored Hidden Wiki (http://zqktlwiuavvvqqt4ybvgvi-7tyo4hjl5xgfuvpdf6otjiycgwqbym2qad.onion/wiki/index.php/Main_Page). Holds a database of links to different TOR websites. You should be suspicious regarding links included in this wiki, as you cannot trust the people who edit it; this is the reason for having the "Uncensored" in its name.

3. Torch (http://xmh57jrknzkhv6y3ls3ubitzfqnkrwxhopf5aygthi7d6rp-lyvk3noyd.onion). TOR search engine claims to index around 1.1 million pages.

4. TorLinks (http://torlinksge6enmcyyuxjpjkoouw4oorgdgeo7ftnq3zodj7g2 zxi3kyd.onion). List popular TOR websites similar to the Hidden Wiki.

5. Duckduckgo (https://duckduckgogg42xjoc72x3sjasowoarfbgcmvfi-maftt6twagswzczad.onion). Search the surface web from within the TOR network.

Summary

Privacy and anonymity are fundamental concepts in cybersecurity; in this chapter, we tried to give a technical overview of how web tracking occurs, parties interested in doing it, and the best method to prevent outsiders from recording our browsing history.

After reading this chapter, you may think that web tracking is 100% evil; however, this is inaccurate. For instance, web tracking is used for many good reasons, such as:

1. Many websites use device fingerprinting to strengthen their authentication systems and prevent unauthorized access to user accounts. For example, an online banking website will capture your current device fingerprint info upon the first login; now, if an intruder successfully compromised your account credential and tried to access your banking account from a different device, the banking application will deny access – as the new device fingerprint is not associated as a trusted device with your account. In such a case, the banking application may ask the user to confirm his identity using another authentication factor (e.g., one-time password).

2. Device fingerprinting can be used to identify device security vulnerabilities (e.g., discover outdated OS and applications), thus recommending updates and patches to client operating systems and installed applications.

3. Digital fingerprinting is used to recognize online bots to prevent them from filling and submitting online forms.

In the second part of this chapter, we differentiate between the three layers of the web and cover the concept of online anonymity and how we can achieve it using VPN and anonymous networks. Online tracking mechanisms can be avoided by using anonymity networks; the two most popular anonymity networks are TOR and I2P. TOR browser allows users to access the TOR darknet and surf the surface web anonymously without being tracked.

In the next chapter, we will cover an important topic that is considered a cornerstone in securing IT systems: cryptography.

Notes

1 Statista, "Facebook's Global Revenue as of 3rd Quarter 2021, by Segment", Novemeber. https://www.statista.com/statistics/277963/facebooks-quarterly-global-revenue-by-segment/
2 Hrw, "US Government Mass Surveillance Isn't 'Secret'", February 12, 2020. https://www.hrw.org/news/2019/09/18/us-government-mass-surveillance-isnt-secret
3 EFF, "Is Every Browser Unique? Results From the Panopticlick Experiment", February 11, 2020. https://www.eff.org/deeplinks/2010/05/every-browser-unique-results-fom-panopticlick
4 Geti2p, "Garlic Routing and 'Garlic' Terminology", January 5, 2022. https://geti2p.net/en/docs/how/garlic-routing

3

Introduction to Cryptography

Introduction

As the world continues to digitize, computer networks have grown exponentially. Organizations of all sizes and across all industries are now building massive computer networks to facilitate their operations and increase productivity. The connection of these networks to the internet has become indispensable in today's information age to foster collaborations with partners, customers, and third-party suppliers.

Nowadays, most organizations' data have become digital, bringing numerous advantages such as speed sharing, unlimited storage capacity, improved accessibility, simultaneous access to data, and reduced chances of missing data. However, this also raises challenges on how to protect this data.

To protect digital data from unauthorized access and maintain the privacy of online communications, cryptography was employed. However, there is an associated risk for every opportunity presented by the information age. For instance, as computer cryptography tools become widely accessible through the internet, threat actors are using them to remain anonymous online and to safeguard their incriminating data against law enforcement.

This chapter will describe the term cryptography, talk about its types, and see how it can secure online communications and digital assets. An important concept associated with modern Encryption is digital authentication. We will describe it, list authentication factors and types, and discuss the most popular digital authentication methods already in use, including the modern one, passwordless authentication. Data security is a broad topic; we will talk about its core component (Data Classification) and see how to secure digital data when it is at rest, in use, or in transit. We will conclude this chapter by talking about the second type of cryptography: Steganography.

What Is Cryptography?

Cryptography is the science, art, and practice of using various methods to keep information and communications private in the presence of a third

DOI: 10.1201/9781003008279-3

party (e.g., the public or other malicious actors). Today, when we talk about cryptography, we usually mention it in the technology context such as securing data transmitted over computer networks or encrypting data stored on storage media such as hard drives or USB pen drives.

Cryptography's name is derived from Ancient Greek, which means "secret writing"; this science has been used to secure private messages since ancient times. Cryptography is used synonymously with "encryption", although using both names interchangeably is not accurate.

Cryptography can be broadly classified into two major segments:

1. Encryption
2. Steganography

Encryption is a cryptography component that most people refer to when talking about cryptography. Encryption works by converting plaintext data into a ciphertext code using a specific cryptographic algorithm. Only people who access the associated decryption key can reverse the process and decrypt the data back to its original state.

The concept of Encryption is too old, and it dates back centuries when some elementary forms of Encryption were used to secure the convey of confidential messages. However, as we move steadily toward digital societies, using Encryption has become vital to ensure the privacy and security of digital data worldwide.

From a cybersecurity perspective, Encryption is used everywhere online, and different entities are already using it to protect data from unauthorized access. For example, governments use Encryption to protect diplomatic communications, military, and classified information. Enterprises use it to protect their trade secrets and official communications; online merchants such as Amazon and eBay use Encryption to secure customer payment info (e.g., customers personal and payment information). Internet messaging (IM) applications such as WhatsApp and Signal use end-to-end Encryption to secure chat messages and reassure users that the message they sent cannot be read by anyone, even the company that made the application. On the other hand, we are seeing increasing awareness among end-users to use Encryption to secure their computing devices (e.g., smartphones, tablets, laptops) and keep their sensitive data private.

Note! Most people still use the term Cryptography to refer exclusively to Encryption.

Steganography is the science of concealing a secret message within what appears to be an ordinary message. In digital Steganography, we use a digital file such as an image, audio, video, compressed file, or even an MS Office document to conceal a secret message within it. The secret message could also be another digital file. This method protects the secret message by hiding it in another file. Steganography is usually combined with Encryption to add additional security to safeguard sensitive data by encrypting the private message before hiding it in the overt file.

Cryptography Algorithms Types

Cryptographic algorithms can be classified using different criteria; the most obvious is categorizing them according to the number of encryption keys used in the encryption/decryption process.

An encryption algorithm needs a key to work. An encryption key is a string of data used along an encryption algorithm to transform the plaintext data into ciphertext and vice versa. The strength of an encryption algorithm depends mainly on the encryption key length (measured in bits). For instance, the bigger the key size, the harder it to crack. However, some weak implementations of encryption algorithms may suffer from bugs and design flaws; this effectively reduces its strength even though a large key is utilized to encrypt the data.

The three main encryption algorithms we will discuss are symmetrical, asymmetrical, and hashes.

Symmetric Cryptography

Also known as secret-key cryptography, in symmetrical Encryption (see Figure 3.1), both sender and receiver use the same key to encrypt and decrypt the message or file. The security of the systems relies on the secrecy

Symmetrical Cryptography

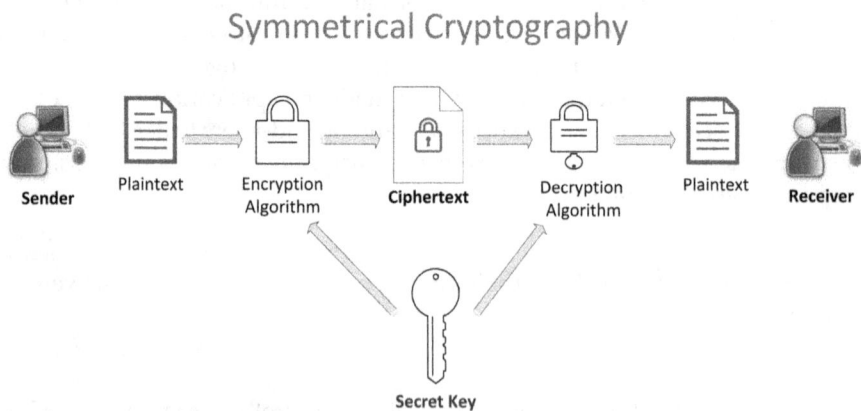

Sender — Plaintext — Encryption Algorithm — Ciphertext — Decryption Algorithm — Plaintext — Receiver — Secret Key

FIGURE 3.1
Symmetric key cryptography systems.

of the encryption key; if this key falls into the wrong hands, the entire system will fail.

Let us give a simple example to demonstrate how secret key cryptography works:

1. Suppose *Randi* wants to send a private message to *Bob*; she uses a shared secret – or a password – to encrypt the message and send the resultant ciphertext to *Bob*.

2. Now, for *Bob* to read the message, he needs to use the same encryption key to decrypt the message again to its readable state.

The most famous secret key cryptography algorithms are Data Encryption Standard (DES) – which has become obsolete now, Triple DES, Advanced Encryption Standard (AES), Blowfish, and Twofish.

Symmetric algorithms are faster than asymmetric algorithms because they consume fewer CPU cycles during the encryption/decryption process; this fact makes symmetric Encryption usually used to protect data at rest (e.g., encrypting hard drive data).

The main disadvantage of the symmetrical algorithm lies in distributing the secret key privately between the communicating parties. To mitigate this obstacle, asymmetrical Encryption was proposed.

Asymmetric Cryptography

Also known as public-key cryptography, in this type, we use two different keys (see Figure 3.2), one for encrypting the data, which is public and can be distributed openly, and the other key, private, for decrypting the data and

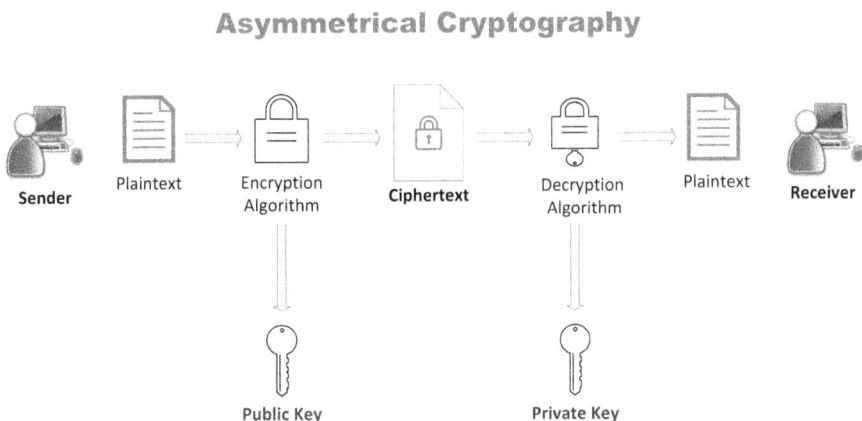

Asymmetrical Cryptography

Sender — Plaintext — Encryption Algorithm — Ciphertext — Decryption Algorithm — Plaintext — Receiver

Public Key   Private Key

FIGURE 3.2
Asymmetric encryption, public-key cryptography.

must remain confidential. Although both keys are mathematically related, you cannot compute one key by knowing the other.

Let us give an example of how asymmetrical Encryption works:

1. Let us assume *Randi* wants to send a private message to *Bob*; she needs to encrypt it using *Bob's* public key, which he already made public.
2. After receiving the encrypted message, *Bob* needs to use his corresponding private key to decrypt the message to its original state.
3. If *Bob* wants to send an encrypted reply to *Randi's* message, he needs to use *Randi's* public key to encrypt the message before sending it.
4. To read the encrypted reply, *Randi* needs to use her corresponding private key to decrypt the message to its original state.

In asymmetrical Encryption, the public key should be publicly available and is used to encrypt a message or verify a sender's Digital Signature, so anyone who wants to send you an encrypted message can use it. The corresponding private key should always remain secret. Otherwise, anyone who has access to this key can impersonate the key owner and send/receive encrypted messages instead of him.

Digital Signature

The digital signature is a technical method to verify (authenticity) the originator of the encrypted message or data and assure that the message has not been tampered with in transit. A digital signature uses asymmetrical algorithms to do its job, as the user needs to use their private key to encrypt the data instead of using the receiver public key. Let us explain how it works using an example.

In our last example of asymmetrical Encryption, we said that if *Randi* wants to send an encrypted message to *Bob*, she needs to use his public key to encrypt the message. Now on the receiver side, *Bob* needs to decrypt the message using his corresponding private key; however, how *can Bob* be assured that this message was originated from *Randi*? What if a malicious party pretending to be *Randi* has sent this message? Here comes the role of the digital signature to verify the authenticity of the claimed sender.

Instead of using *Bob's* public key to encrypt the message, *Randi* uses her private key to encrypt the message before sending it to *Bob*, now. For *Bob* to read the encrypted message, he will use *Randi* public key to decrypt the message. When the message successfully decrypted, *Bob* can be confident that this message is originated from *Randi* because it works with *Randi* public key and *Randi* is the only person who has access to her private key that she used to encrypt the message (see Figure 3.3).

Digital Signature

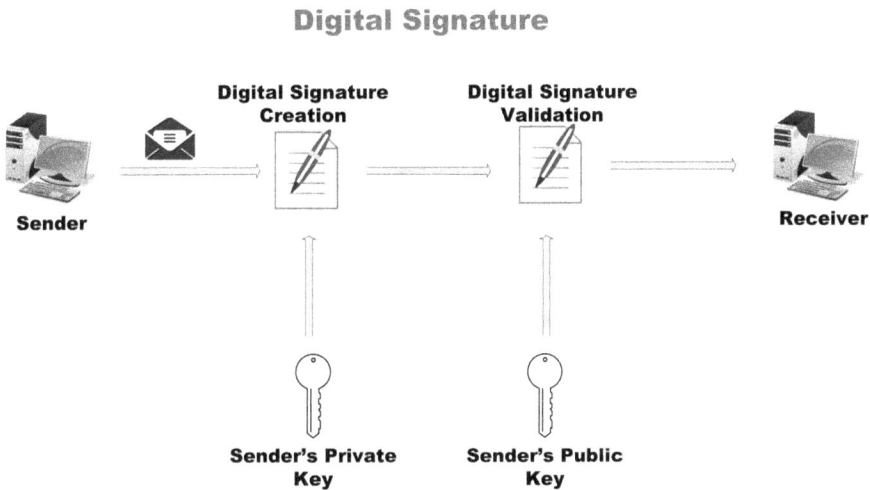

FIGURE 3.3
Digital signature.

Cryptographic Hash

Hashing is the third type of cryptographic algorithm. It is a one-way process designed to be irreversible (see Figure 3.4). Hashing works by inputting a file or any form of data into a hashing algorithm to generate a fixed-size string value. For example, if you create a hash value for a data set, you cannot reverse the process and generate that data again from the hash value. The other types of cryptography (symmetrical and asymmetrical) are a two-way process; hence, they are reversible (in Encryption, we can encrypt and decrypt data using a specific encryption key).

The resultant value from the hashing process is almost unique to the input data; it is extremely unlikely that two different inputs can produce the same hash value. Old cryptographic hash algorithms (such as SHA-1) – which use a short size encryption key – are vulnerable to Hash Collison, which occurs when two different data inputs produce the same hash value using the same hashing algorithm. Although the chances of happening a Hash Collison are very rare, it is advisable to use modern hash algorithms that use a long key size to mitigate this possibility.

Hashing is used widely in different applications to secure data and verify its integrity. For example, when you sign up for most online services, your password will be hashed first using a hashing algorithm and then stored in the credential database. When you want to log in to your account again, the authentication system will convert the entered password into a hash and compare it with the one stored in the credential database. This effectively keeps users' cleartext passwords safe if an intruder gains access to the credential database.

How Cryptographic Hash works

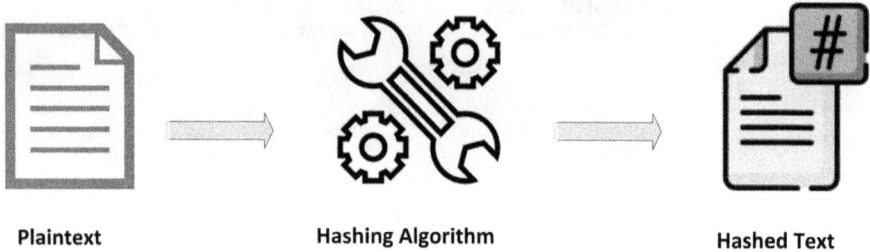

Plaintext   Hashing Algorithm   Hashed Text

FIGURE 3.4
Encryption hash.

Hashing is used to verify the integrity of data. For instance, digital forensics examiners compute the acquired digital evidence image hash value and compare it with the source media hash value to ensure it is an exact match. Hashing in this scenario is also named *Digital Fingerprint* because it uniquely identifies the subject file. Software downloaded from the internet can also be verified using hashing by comparing the file hash value posted by its owner online with the computed hash value of the file after downloading it. If both hash values match, the downloaded file is not corrupted and has not been tampered with during the download process.

There are different types of hashing algorithms; as we said before, we should use the one with the longest key size; the most popular and secure ones are SHA3 and Whirlpool.

How to Calculate File's Hash Values

We can calculate the hash value of any file using a built-in hash calculator available in major operating systems (Windows, macOS, and Linux). Let us experiment with doing this under Windows 10.

1. Launch a PowerShell window by pressing **Windows key + X** buttons to open the Quick Access Menu.
2. Click on Windows PowerShell (Admin).
3. Type **Get-FileHash** followed by the path to the file you want to calculate its hash value (see Figure 3.5).
4. This command will automatically calculate the hash value using SHA-256 cryptographic algorithm. However, you can specify a different hashing algorithm using the Algorithm command followed by the algorithm name.

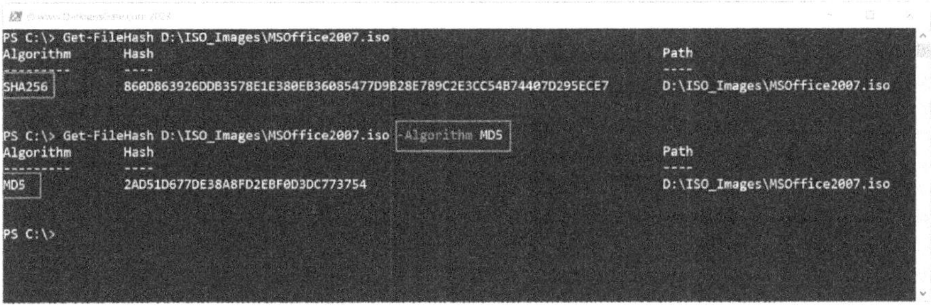

FIGURE 3.5
Using PowerShell under Windows 10 to calculate files hash values.

There are many third-party tools to calculate files hash values; the following are the most popular:

1. QuickHash GUI (http://www.quickhash-gui.org)
2. HashCalc (https://www.slavasoft.com/hashcalc)
3. Hash Tool (https://www.digitalvolcano.co.uk/hash.html)

Authentication & Authorization Concept

Digital authentication (also known as electronic authentication) is the process of establishing the identity of a legal person against an authentication mechanism (information or a physical system) to enable access to restricted resources. Digital authentication is also used in many instances to confirm a person's identity, such as using digital signatures or confirming the authenticity of a website by using digital certificates.

People can prove their identities using different official papers, such as a national ID, a driving license, or a passport, in the real world. However, things become more complex online, as people need to prove their identities remotely across computer networks, and here comes the role of e-authentication to solve this obstacle.

Digital authentication is not restricted to people. For instance, workstations, servers, networking devices, network services, and processes utilize authentication to verify their identities when communicating across computer networks.

Authentication Factors

When authenticating users online, we can differentiate between three main types of authentication factors: something you know, something you have, and something you are. Nevertheless, many digital authentication mechanisms use more than one authentication factor to authenticate users for added security (e.g., using a password (first factor) and a credit card (second factor) to withdraw cash from ATMs).

Something you know: This includes regular passwords, PIN, passphrase, or anything a user must remember (e.g., answer to a security question). Password-only authentication systems are still widely prevalent.

Something you have: This includes anything a user possesses in hand and uses it to access the restricted resources. Examples include hardware or software tokens to generate a one-time password (OTP), mobile phones to receive a temporary password via SMS, credit cards, and access cards.

Something you are: This is the most expensive authentication model because it needs special devices to recognize a user biological and behavioral characteristics. In this type, users supply their biometric data (also known as biometrics) such as fingerprint, DNA matching, voice patterns, face recognition, palm print, hand geometry, iris scan, or behavioral biometrics for authentication.

With the advance of computing technology, two additional authentication factors have been used in modern authentication systems: *somewhere you are* and *something you do*.

Somewhere you are: In this method, additional technical information about the computing device used to conduct the authentication is examined, such as device IP address (to identify user geolocation data), Media Access Control (MAC) address in addition to user's digital device fingerprint (discussed in the previous chapter).

Note! What is the MAC address?

Each network card has a unique number (12-digit hexadecimal number) called the physical address or MAC in short. A MAC filter is commonly used in enterprise network environments to govern access to network resources by allowing only authorized devices to gain access to protected resources based on their MAC address.

It is worth noting that a single device can have two MAC addresses. For example, laptops with a Wi-Fi card have a MAC for the Ethernet port and another for the Wi-Fi card.

Something you do: This is a new authentication type commonly used in mobile devices with touch screens. It works by prompting users to draw something on their computing device screen to gain access. A clear example is unlocking mobile devices' screens – and some applications – by swiping or drawing a specific gesture on the screen (see Figure 3.6).

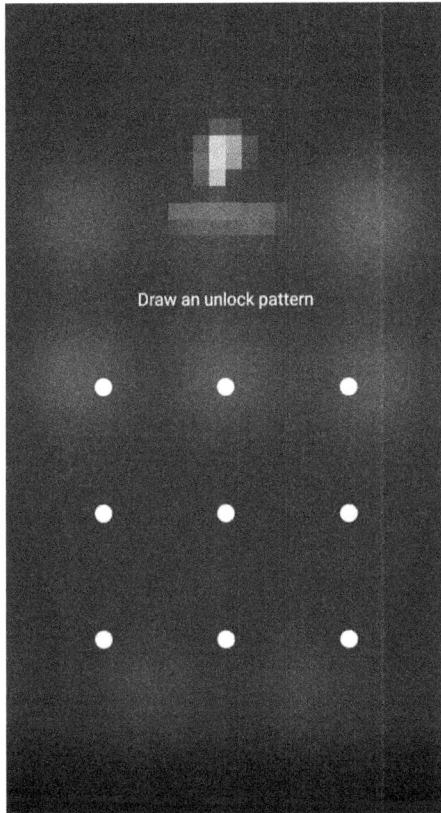

FIGURE 3.6
Draw a pattern to unlock an app on a smartphone screen.

Most authentication systems are still using passwords primarily to authenticate online users. However, more and more systems – and people – are becoming aware of using more than one authentication factor to prove their identity. For example, supplying a password and receiving an OTP via an SMS message can effectively deter many attacks aiming to compromise users' credentials such as brute-force and credentials stuffing attacks.

Authentication Types

Digital authentication can be categorized according to the number of authentication factors in use:

Single-factor Authentication

Also known as "Primary Authentication", in this type, only one authentication factor – such as a password or PIN – is used to authenticate users.

Single-factor is the simplest type, and it is considered the most straightfor-ward authentication form for users. However, it comes with poor security because malicious actors can access other users' accounts by just knowing their passwords.

Two-factor Authentication

This is a more secure option than single-factor. In this type, two authentication factors are utilized to verify a user. For example, Smartphones equipped with GPS can identify user location and use it as the first factor, while the second authentication factor can be the traditional password provided by the user.

Multi-factor Authentication

This is the most secure authentication mechanism and is usually used to pro-tect high-value resources. In this type, the authentication system leverages more than two authentication factors to authenticate a user. For example, the follow-ing factors can be utilized to grant access to a particular restricted resource:

A. A regular password.

B. An OTP is sent via SMS to a user cell phone.

C. User fingerprint or FaceID (biometric data).

D. User location data (GPS coordination) taken from user GPS-capable device that donates the location where the authentication attempt is made.

Authentication Methods

Various authentication methods are employed to verify the identity of some-one (a user or device); the following are the most popular ones.

Passwords

This is the simplest and oldest method used for digital authentication; in this type, users provide their username and password to gain access. Users' passwords are – generally – stored hashed in the credential database to avoid revealing them if an adversary gains access to the database. Although it is considered weak and vulnerable to many cyber-attacks, password-based authentication is still widely used – as the default scheme – to protect user accounts in major social networking platforms and popular shopping sites.

Biometrics

This type is considered the most secure authentication method because it relies on users' unique biological characteristics, which cannot be falsified easily.

Years ago, biometric authentication systems were expensive and difficult to deploy. They were usually used to protect high-value assets, such as controlling access to digital forensic lab rooms and other sensitive places like airports and military bases. However, with the advance of computing technology, biometric authentication features have become incorporated into end-user devices, such as smartphones (e.g., Apple Touch and Face ID) and other high-end personal computers. Today, more people are willing to use it as a convenient and secure alternative to password-only-based authentication (e.g., Windows 10 provides biometric authentication via the HELLO App – Introduced with Windows 10 version 1507).

The physical characteristics of humans commonly used for biometric authentication are fingerprint scan, facial recognition, voice identification, eye scan, and hand geometry.

> **Note!** Biometric authentication is more secure; still, it is not a foolproof solution. For instance, Hackers spoofed the facial recognition feature by creating a 3D model of the target user head based on pictures taken from publicly available sources, such as Facebook and Instagram. Another attack works by utilizing AI to generate an identical fingerprint to fool fingerprint scanners available in smartphones.[1]

Tokens

A security token (also known as an authentication token) is commonly used to enhance the security of the traditional password-only authentication method by adding a second authentication factor to the process. For example, a user needs to provide a security token in the form of a USB device that plugs into a USB port, in addition to a regular password.

Security token comes in two forms: Software and Hardware-based, and they differ in where the application is stored. Hardware security tokens are portable devices that hold a user unique identifier and are used to gain access to an electronically restricted resource. Most popular hardware token types come as smart cards or mimic USB keys (commonly called a dongle). Other advanced types contain an LCD to show the generated OTP, or have a keypad for entering passwords, a biometric reader such as a finger scanner, or come supplied with Bluetooth or another near-field communication (NFC), radio-frequency identification (RFID) to authenticate. Hardware tokens come with some disadvantages. For example, they are easy to lose, relatively expensive, and vulnerable to theft (although some hard tokens come protected with a password if they fall into the wrong hand).

Software tokens are programs installed on end-user personal devices such as computers or smartphones and do not require purchasing additional devices. Software token performs the same duties as the hardware token; however, using it is easier as it does not occupy any physical space. We can duplicate it easily, cost nothing, and it is not vulnerable to physical theft. Software tokens are often installed on smartphone users (this mobile-based authentication approach is called "Phone-as-a-Token"). Despite all the advantages associated with the software token, it still suffers from malware threats like any other computer program. For instance, if the computing device holding the token software gets exploited by a malicious actor, this actor can access the digital certificate or the unique identifier stored in the token application.

Multi-factor Authentication

We already talked about this concept; in multi-factor authentication (MFA), we use two (2FA) or more verification factors to verify user identity. Most people use this form of authentication when using a bank ATM to withdraw money. For example, a customer needs to insert his/her credit card in the ATM and then provide a PIN. Another example is an online banking system requiring users to provide an OTP sent to their mobile phone via SMS or email after providing the correct typical authentication credentials (username and password).

Public-key Authentication

In this system, users verify their identity using a cryptographic key rather than a password. This system works as follows: each user has a key pair consisting of a public key (publicly available and usually stored in the authentication server) and a private key that must be kept secret. The authentication server holds a list of users and the corresponding public key for each one.

When a user wants to access the restricted resources, the server asks them to provide the private key corresponding to the stored public key. The server will make the required calculation, and if the provided private key is the correct pair of the associated public key, access is granted.

Private-key Authentication

This method is similar to public-key authentication; however, a user has only one key used for authentication instead of having two keys. This method depends on the secrecy of the private key, which is stored in two locations: the authentication system and the user.

This method works as follows:

1. A user's private key is already stored on the authentication server.
2. The authentication server sends a random message when a user wants to log in to the protected resources. The user encrypts this message using their private key and resends it back to the server. Now, the server compares the encrypted message sent by the user with the one encrypted using the user's private key stored on the server. If there is a match, the user is grant access.

Compared with public-key authentication, this method is less secure because the user private key is stored in two locations. If adversaries exploit the authentication server, they can grant access to all stored user accounts. Some implementations use a static password and the private key to strengthen this method, making it a two-factor authentication (2FA) system.

Digital Identity Authentication

This is a modern authentication method commonly used in modern web applications. In this method, we are not relying on static passwords or security tokens. Instead, the authentication mechanism stores anonymous information (non-Personally Identifiable Information) about each user computing device (laptop, smartphone, tablet). Such as device technical specifications (digital fingerprint), a user location (taken from the IP address), and other online behavioral histories. This info is then associated with additional offline info provided by the user when making purchases or accessing other online accounts such as billing address, phone number, email address, and shipping information.

Suppose a user tries to access from a different device to their online account. In that case, the authentication system will inform the user and deny access because the user is using an unrecognized device. In that case, the user needs to prove his identity in another way, such as entering a PIN sent to the user's phone or email address.

Passwordless Authentication

The first usage of a computer password date back more than 55 years ago, when the researchers at the Massachusetts Institute of Technology in mid-1960 invented a massive time-sharing computer called CTSS. Multiple persons need to use the CTSS machine, to separate their private files, they came up with an idea to have a password that distinguishes each individual user.[2]

At that time, password security was not an issue, because its usage was mainly to secure an isolated machine (not connected with other devices) used by many persons, as with the case of CTSS. However, after the proliferation of the internet and the increased number of internet users worldwide, compromised credentials have become the primary source used by threat actors to hack into organizations' IT systems and networks. According to *Verizon 2021 Data Breach Investigations Report*, 61 percent of breaches were caused by breached credentials.[3]

There are many cyberattacks that depend entirely on attacking passwords to gain unauthorized access, such as phishing, brute-force, dictionary attack, credentials stuffing, password spraying, and malware keyloggers. This fact has encouraged the computer security industry to move toward a passwordless authentication schema to mitigate the different security vulnerabilities associated with using password-only systems.

Before talking about passwordless in details, let us first discuss why password-only authentication systems are weak.

Why Passwords Are Bad?

Regular passwords are bad for the following reasons:

1. People tend to use the same password to secure more than one account or application. This is a very bad security habit, as compromising one account credentials can lead to compromising all accounts belonging to the same person. A recent research by Yubico,[4] found that 54% of employees admitted to use the same passwords across multiple work accounts.

2. People tend to use easy-to-remember passwords. This makes guessing passwords easy using automated brute-force attack tools.

3. Until now, many people write their passwords down on a piece of paper or keep them inside an unprotected Excel file.

4. Traditional passwords are shareable; sometimes user share their passwords using insecure channels, such as social media messages or via email making them susceptible to fall in the wrong hands.

Despite the numerous security problems associated with using passwords, they are still widely used throughout all industries. A study by *Transmitsecurity*[5] about passwords impact on customers experience found the following:

- 55% of consumers have stopped using a website because the login process was too complex
- 87.5% of consumers have found themselves locked out of an online account after many failed login attempts
- 92% of users will leave a website instead of recovering or resetting their login credentials

As we saw from *Transmitsecurity* study, passwords affect user experience in different ways, and getting rid of them can improve users experience and increase authentication security, and this what passwordless authentication is trying to achieve.

What Is Passwordless?

As its name implies, passwordless authentication is the process of verifying users' identities without providing any password or any memorized secret (such as a passphrase), such as:

1. Using users biometric signature.

2. Receiving a link to the user registered email address containing a one-time link.

3. Having a hardware device, such as a hardware token or other supported devices, such as a smartphone or tablet, that contains an authentication application installed on them to generate an OTP. The OTP can also be sent via SMS to user registered phone number.

4. Via a third-party identity provider, such as using Facebook, LinkedIn, or Google to authenticate against supported services.

5. Using a persistent cookie.

Passwordless authentication provides numerous benefits for both end-users and businesses:

1. Users will not worry about remembering many passwords to secure their digital accounts.

2. Eliminate various cyberattacks associated with traditional passwords, especially phishing and brute-force attacks.

3. Enhance users experience as there is no need to create complex passwords and to memorize them.

4. Eliminate the hassle on the IT support staff resulted from users requests to reset their forgetting passwords.

5. Businesses will avoid the overhead of storing and maintaining large numbers of users' passwords.

Passwordless Authentication Types

There are different types of passwordless authentication; the following list the most used ones:

1. *Biometric authentication*: This type has become widely used in modern mobile devices. Windows Hello app and Apple Face ID are two examples. Biometric is commonly used to authenticate users to their mobile devices, and is seldom used to enter specific online resources.

2. *Hardware security tokens*: This type requires the user to possess a device, such as a security key or USB dongle containing a user private key.

3. *An OTP*: OTP is the most used method to authenticate users without a password. In this type, users provide their mobile phone number during the authentication, and an SMS message containing a temporary code valid for a short time. Sometimes, a notification generated from an installed application on user smartphone is used instead of SMS. *Google Authenticator* is an example.

4. *Magic Link or Code*: In this type, users provide their email address to the authentication service, and an email is then sent containing a magic link to access the restricted resources without providing a password. Slack is an example of an online service that support this type of authentication.

5. Third-party authentication: In this type, users use a third-party service such as Google, Facebook, LinkedIn, Twitter, or Apple to prove their identity. The Medium website uses such a passwordless authentication mechanism to facilitate user login.

Data Security Through Encryption

In today's digital age, data has become the lifeblood of organizations, and keeping it secure is the primary task of every cybersecurity professional. Data security is the practice of implementing various techniques and security access controls to protect sensitive data from unauthorized access, modification, destruction, or usage. The ultimate goal of data security is to ensure data confidentiality, integrity, and availability, and this can be achieved by implementing different protective measures, such as:

- Data Encryption
- Data Hashing
- Data redaction (masking)
- Implementing different access controls to safeguard data (both logically and physically)

As digital transformation continues to accelerate, the adoption of cloud computing is increasing at an explosive rate. According to Gartner,[6] by 2025, the enterprise IT spending on public cloud computing will exceed their spending on traditional IT. The massive shift to the cloud will result in moving a large portion of organizations' apps and data to the cloud, which makes securing cloud assets a major concern for organizations in the near future.

Protecting data, both on-premise and in the cloud, is essential to prevent data breaches and to achieve regulatory compliance. Different security controls must be enforced to protect data (we will cover them thoroughly in the coming chapters, both on endpoint devices and on enterprise networks), however, the most important measure to secure data is through implementing Encryption and storing the encryption keys securely.

Encryption ensures data remain confidential when storing or transmitting them across insecure mediums, such as the internet. Encryption also ensures data integrity by using hash algorithms and message digests; this assures the receiver that the received data has not been tampered with during transit. Despite the great benefits of using Encryption to secure data, it is still not widely spread in the enterprise world for the following reasons:

1. A part of organizations data that contains sensitive data is unstructured. For instance, unstructured data such as rich media, audio, geo-spatial data, documents (MS Office files, PDF, Text, and drawing

files), emails, invoices, Internet of Things (IoT) sensor data, and many unstructured data. Encrypting different file types is not an easy task, especially when you have a large volume of unstructured datasets to encrypt.

2. Accessing encrypted data is not easy for the average user, especially when using another device to open the encrypted content. For example, most organizations automatically allow an employee's work device to open encrypted content; however, problems arise when moving encrypted content to another device. For instance, the user needs to provide the required security certificate and password and install the required software to decrypt the data.

3. Lack of guidelines on which data to encrypt. For instance, most organizations do not have a clear data classification policy on which data they should protect using Encryption. Many organizations apply data encryption to stored data (or data at rest), while leaving data transmitted between devices or across the internet unprotected.

4. Difficult to integrate Encryption with existing business software. For example, organizations use different applications, such as ERP and HR apps, to process sensitive information; integrating Encryption may break the workflow of these applications and introduce different issues to the normal work processes.

To use Encryption correctly and enjoy its benefits while mitigating most issues that prevent organizations from adopting it widely, it is essential to first identify the data that you need to protect, and this what we are going to cover next.

Data Classification

In the information security context, data classification refers to the process of analyzing structured and unstructured data and categorizing it into groups, or categories, according to some criteria, such as content type, file type, and its sensitivity level. The sensitivity level can be determined by measuring the impact of revealing, losing, or destroying such data by an adversary. The classification of data brings two major benefits for organizations:

1. It helps organizations know which data they should protect to meet the enforced legal, and regulatory compliance requirements, such as: (e.g., Health Insurance Portability and Accountability Act (HIPAA), Sarbanes-Oxley Act (public companies) and the General Data Protection Regulation (GDPR) that require enforcing specific security requirements when handling the PII of EU citizens data.

2. The classification of data helps organizations understand the required security controls and appropriate encryption scheme needed for each data category.

There are different data sensitivity levels, or labels, used worldwide. Some organizations have their own naming; however, regardless of the used naming, the following general data sensitivity levels are the most used:

- **Restricted** or highly sensitive information: Such information is only accessed upon need. Examples of such data: trade secrets, personally identifiable information (PII) of customers/employees, PHI, and financial information.
- **Confidential**: Such information is used within an organization to support its business operations, and should be available for certain people only. Examples include: business and marketing plans, contracts with third-party providers, and suppliers.
- **Internal**: This type of data is intended for internal usage by organization employees. Examples of such information include: training guides, usage manuals, and company directories.
- **Public**: As its name implies, public data can be shared freely across all mediums. Examples include: information posted on a website and company brochures.

Classifying your data will help your organization focus on protecting the most sensitive assets and avoid wasting money and resources to protect unimportant data.

> Note! Data classification is considered a part of the Data Security plan, which is concerned in evaluating security risks associated with storing any type of data and suggesting countermeasures for protecting it.

Digital data can be found in three different states. In the following section, we will talk about these states and suggest the best security measures to protect data within each one.

States of Digital Data

Digital data exist in computing systems using any of the following three states:

1. Data At Rest
2. Data In Transit
3. Data In Processing

Most data change states frequently, for example, customers data may need to be processed using some analytical software to generate sales forecasts. However, some types of data remain in one state and do not change for a long time, such as data stored on backup tapes.

Understanding the different states, or modes, of data will allow organizations to plan their protective measures of sensitive data more efficiently.

Data At Rest

Data At Rest, refer to data stored in persistent storage device, either on-premises or using a cloud storage service. There are different examples of data at rest, such as data stored on employee's laptop, USB flash drivers, external storage device, data stored in a storage area network (SAN), databases, spreadsheets, files backup tapes, archived data, and any inactive digital data that is not currently being transmitted via computer networks or processed in any application or CPU.

Data At Rest may frequently get fetched for manipulation by a user or an application; however, a large volume of data-at-rest remains inactive or accessed for long time, such as archived and backup data. Data at rest is considered less vulnerable to cyberattacks compared with data in transit; however, threat actors prefer to target this type of data because of its importance and the ease of exploiting it later. For instance, data stored at rest is rich data and commonly contains structured data, such as PII of customers, vendors, suppliers, protected health information (PHI), financial transaction history, intellectual property (IP), and trade secrets. This type of data can be exploited directly for commercial gain, while capturing data flowing across network is a daunting task and requires time and resources to analyze and decrypt, this makes Data-At-Rest more valuable than data-in-transit in the attackers viewpoint.

When data is at rest, it is protected using traditional defenses such as firewalls and antivirus programs; however, these solutions are not impenetrable. Organizations need additional layers of defense to protect data at rest if threat actors successfully penetrate the network and gain access to the data repository.

We can implement various measures to secure data at rest, such as:

- Deploy a Data Loss Prevention (DLP) solution to prevent data exfiltration. DLP will block USB, smartphone, or any other external device from accessing the computing devices containing sensitive data.
- Use the Tokenization method to hide sensitive data. A token is the piece of nonsensitive data used to represent a sensitive piece of information, such as a customer credit card number.

- Use Masking to hide part of the sensitive data using an asterisk. For example, the sensitive part of the credit card number can be masked while leaving the remaining unsensitive part unmasked (see Figure 3.7).
- Enforce physical security access controls to prevent accessing the storage devices or stealing them.
- Utilizing MFA to restrict users access to data.
- Dividing data into chunks and storing it in separate physical locations, this prevents threat actors from gaining complete information if successfully compromised one location.
- Store sensitive data offline (e.g., offline storage devices, such as tapes or disconnecting the storage server from the internet).

In addition to all the mentioned above protective measures, the best technical solution to keep data at rest safe is to use Encryption. Data At Rest are commonly encrypted using symmetrical Encryption because it is very fast, as it requires few CPU cycles to process, in addition to being easy to implement since there is only one key used for both Encryption and decryption.

> Note! We should almost always encrypt Data At Rest. In many data privacy regulations, if Data At Rest has been compromised while it was encrypted, and the attackers failed to gain access to the decryption keys, then there is no need to report the incident publicly.

Data In Transit

Data In Transit is also known as data in motion; this data is in an active state; hence, it travels from one device to another via computer networks, which include both internal networks or the public internet. Example of data in transit:

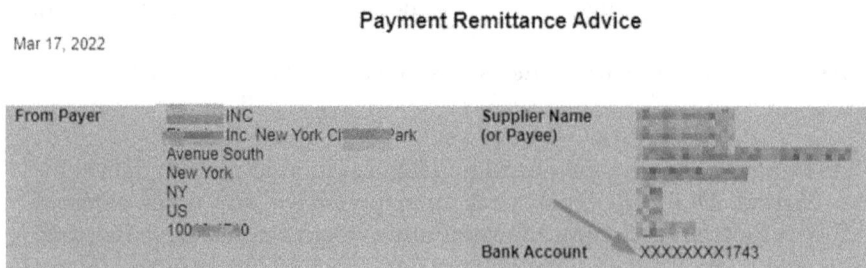

Payment Remittance Advice

Mar 17, 2022

From Payer	INC	Supplier Name
	Inc. New York Cl____Park	(or Payee)
	Avenue South	
	New York	
	NY	
	US	
	100_____0	
		Bank Account XXXXXXXX1743

FIGURE 3.7
Use the masking technique to redact the sensitive part of the content.

email messages, IM, and sharing files across an organization intranet. Moving data from cloud storage to on-premise servers is also considered data in motion. Data in motion can be traveled via wired cables or via wireless connection.

Since the creation of the internet, securing data in transit was a major concern for industry and governments, and this was the driving force to develop encryption techniques to protect data while traveling across unsecured networks such as the internet. Data in transit is the most attacked by cybercriminals; anything that goes online or is delivered through computer networks can be susceptible to interception by different threat actors. For this reason, it is necessary to encrypt all traffic traveling across computer networks (both intranet and the internet).

When talking about protecting data in motion, we mainly mean protecting data transmission through email and cloud interactions. For instance, when sending an email, it travels across many points online before reaching the final destination. The best method to secure email messages and their attachments is to utilize Encryption. There are also different security protocols used to protect data at transit when working online (e.g., accessing cloud storage) such as Transport Layer Security (TLS) or its predecessor, Secure Socket Layer (SSL) in addition to VPN and or IPsec.

To protect data in motion, the following security measures should be followed:

- Data can be encrypted first, while it still at rest, before transferring it to another location via computer networks,
- Creating a secure tunnel, when moving data between two destinations via the internet or intranet, and moving the data through it. For example, VPN creates an encrypted tunnel that can be used to transfer data securely via the internet.
- In an enterprise environment, it is essential to prevent employees from using alternative channels other than the one approved by the IT team, to upload data to cloud storage. For example, using Dropbox and Google Drive by employees should not be allowed when handling organization's sensitive data without supervision from the IT security team. Because using free services to upload data to the cloud does not always meet the data protection requirements imposed by organizations, which makes data uploaded via these methods more subject to security breaches and data loss.

Data In Use

Data In Use refers to the data fetched into the RAM for processing by an application, system, or database. Examples of data in use include analytical software bringing and analyzing previous customers' purchases from a database to create a profile for each one to target them with customized advertisements. Another example is generating a report of all online banking transactions of a specific bank client.

Protecting this type of data is a challenging task because data in use can be accessed using different methods based on how each organization processes data. In the absence of Encryption (data should be decrypted first before a user or application can use it), securing data in use require implementing strong user authentication within an organization to ensure only privileged can access the data. Organizations should also keep sensitive data off employee's portable devices, so access can be removed instantly if an employee device is stolen or lost. Legal control is also necessary (asking employees to sign a non-disclosure agreement) to prevent employees from revealing their organization's sensitive data to unauthorized people.

Using Encryption to Protect Data throughout the Entire Lifecycle Phases

Based on our discussion of the three states of digital data, we can conclude with the following.

Encryption is still the most protective measure to prevent unauthorized access to sensitive data; however, Encryption can only protect data when it is at rest and in transit. This leaves sensitive data exposed to different vulnerabilities when it is processed on endpoint devices or by cloud applications. To mitigate this problem, In-Use Encryption was invented.

Defining In-Use Encryption

In-Use Encryption is a new scheme that aims to encrypt data while it is under processing. This efficiently prevents threat actors from intercepting data while processing it on-premise or in the cloud, and ensures data is protected across the entire lifecycle phases (at rest, in transit, or in use).

In-Use Encryption is performed using Homomorphic Encryption, which is a modern encryption algorithm that allows performing computations on ciphertext (encrypted data) as if it were performed on plaintext data.

What Data Should Your Organization Encrypt?

This brings us to the major question: should we encrypt all kinds of data? The answer depends on the type of data each organization handles. In general, all regulated data must be encrypted and secured properly, such as:

A. *Personally Identifiable Information (PII)*: This includes any information that can be used on its own to identify a person. Such as: Full name, Social security number, Date of birth, email, phone and fax number, mail address, passport, and national ID's.

B. *Protected Health Information (PHI)*: Patients health information.

C. *Customers Financial Information*: Such as customers credit card numbers, payment histories, and financial statements.

Other information that must be encrypted includes:

Confidential Business & Intellectual Property: This includes company expansion and marketing plans, product design documents, development and research data, financial performance of a company, and any other valuable information that effects on company work and reputation.

Steganography

Steganography is the science of concealing a secret message inside what is looking an ordinary file. It is the second type of Cryptography in addition to Encryption. This science is too old. For instance, the term Steganography was first used in 1499 AD by a German scholar named *Johannes Trithemius* in his three-volume work titled *Stegographia*. His book included different techniques to conceal secret messages in text.

In today's digital age, Steganography is commonly used in conjunction with Encryption; for example, we encrypt the secret message (such as text content) first before hiding it within the overt file, which could be an image, video, audio, or another text file.

Some may argue that Encryption is still more helpful in concealing secret information, especially if this information is very sensitive; however, this is not always applicable. For instance, sometimes, we want to deliver a secret message online without grabbing the attention of automatic monitoring solutions installed by ISPs, Governments, and other well-funded entities. Monitoring solutions can detect the presence of encrypted data when scanning network traffic; however, detecting Steganography is almost impossible because we are hiding our message in plain sight without changing the overt file structure (in many instances).

Steganography Types

Different types of Steganography exist, such as:

1. Linguistic – conceal messages in written neutral language. For example, using symbols and signs to hide the secret message.

2. Technical – such as invisible ink

3. Digital – conceal a secret message within digital files

This chapter will focus on the digital Steganography types because it has its use cases in cybersecurity.

Digital Steganography

In digital Steganography, we conceal a secret message within a digital file. Almost all digital file types can be used to hide our secret message; however, the common file types used as overt files or carriers of the secret messages are:

- MS Office documents and PDF files
- Image files
- Video files
- Audio files
- Networking protocols

How Is Steganography Implemented Technically?

Steganography is implemented in digital files using three techniques:

Adding Bits to a File

The secret message is inserted in the header or after the overt digital file's end-of-file (EOF) marker. For example, in many image file types, the header of the image can contain more information than the image metadata information (size, resolution). Confidential data can be inserted there without changing the overt file quality.

Each file has an end-of-file (EOF) marker that signals the end of the file. After this marker or tag, we can conceal our secret message safely because the application used to read the file will not read anything after the EOF marker. In Figure 3.8, we hide a secret message after the EOF of a WordPad file (RTF file type).

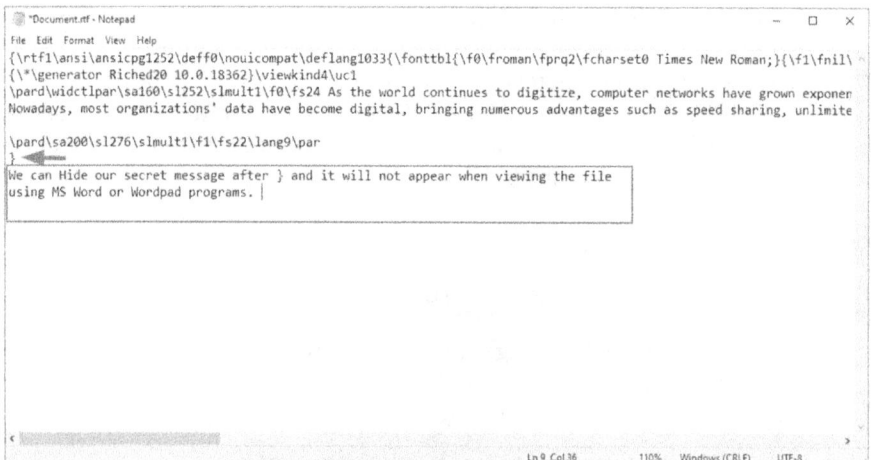

FIGURE 3.8
Hiding secret message in an RTF file after the end of file marker – You need to edit the RTF file using Windows Notepad to insert/read the hidden message.

Using the Least Significant Bit

We are not adding new bits to the overt file in this method. Instead, we substitute one bit from the original (overt file) with one bit from the covert information. Changing the least significant bit will not affect the quality of the overt file. It would not increase the original size of the overt file, comparable with the "Adding bit to a file" method, which will increase the overt file size if the size of the secret file is large.

Generation

In this method, we will not insert or substitute bits in the overt file to conceal our secret message; instead, we will generate a new overt file based on the secret message we want to conceal. A good example of this technique is using SPAM emails to hide the secret message. Spammimic (https://www.spammimic.com/index.cgi) is a website that allows users to write a secret message by generating a spam email containing a secret message.

Steganography Protocols Types

Digital Steganography is implemented using any of the following three protocols:

Pure Steganography

This is the simplest method; Steganography is implemented without using a stego-key. For instance, in our previous example about hiding in an RTF file, we concealed a secret message within an RTF file using pure Steganography. The disadvantage of this method is that threat actors can easily reveal the secret message if they suspect something is hiding within the overt file.

Private Key Steganography

In this method, the secret information (which could be another digital file) is inserted within an overt file (any supported digital file) using a stego-key. The overt file is now sent to the recipient, who needs to have the same stego-key to uncover the hidden message. The problem with this technique is that the stego-key should be exchanged in advance between the communicating parties, which makes it susceptible to interception by threat actors – if exchanged over the internet.

Public Key Steganography

This technique is similar to public-key Encryption. The sender will conceal the secret message using the receiver's public stego-key and send it to them.

Now, the receiver will use their private stego-key to read the secret message. This method is more secure because it does not require exchanging stego-key in advance between the communicating parties.

A significant disadvantage of the last two methods is that concealing a secret message involves using a specific steganography embedding algorithm signatures that can be detected using automated monitoring solutions. Monitoring tools can detect the presence of a secret message hidden in a particular overt file.

Steganography According to Host File Type

Digital Steganography can be grouped using different criteria; the most common one is categorizing it according to the digital file type (overt file) used to conceal the secret message.

Text

We use text to conceal the secret message within it. For example, the "hidden text" feature in MS Word allows hiding text within the document. The text steganography method is not widely used because of the limited volume of secret text that we can hide.

Image

Image steganography has become the most used technique to conceal secret messages because of the widespread usage of digital images online. This technique works by embedding a secret (sometimes encrypted) message into an image file using a predefined stereographic algorithm. The resultant file is named a stego-image. When the receiver gets the stego-image, it will use a shared secret to retrieve the secret message from the overt image file.

Audio/Video

Audio Steganography exploits the human auditory system; for instance, the human ear can recognize noise in the audible frequency range between 20 Hz and 20KHz. Audio Steganography works by embedding the secret message bits within the audio signals without noticing this by the human ear. This is because the human ear cannot listen to low tone frequency in the presence of a higher frequency.

Video steganography works by using images and audio techniques to conceal secret messages. Videos are composed of frames of audio and image files run in a continuous stream. Confidential data can be hidden in this sequence of images/audio files.

Network

Network steganography conceals secret messages by hiding them within some networking protocol headers. For example, we can hide secret messages by exploiting the unused space in IPv4, IPv6, TCP, or UDP headers. Another technique is making a secret message to look like ordinary traffic (e.g., making it similar to HTML code) and sending it through port 80.

Summary

In today's information age, every organization needs to protect its precious data assets, and encryption remains the primary method used to secure sensitive data in its three states: At rest, in transit, and in motion. In this chapter, we talked about cryptography and differentiated between its different types.

The concept of digital authentication is a cornerstone in computer security; we discussed the difference between authentication and authorization, talked about different authentication factors and types, and discussed in some detail the main authentication methods already in use.

Steganography is the art and science of hiding secret messages within an ordinary-looking file. We talked briefly about Steganography, its types, and the main techniques used to embed secret messages within digital files

This was our last chapter for this part. In the next chapter, we will talk about different cybersecurity threats and discuss the most prominent attack vectors employed by cybercriminals to launch their malicious attacks.

Notes

1 Fortune, "Artificial Intelligence Is Giving Rise to Fake Fingerprints. Here's Why You Should Be Worried", Accessed 2025-04-02. https://fortune.com/2018/11/28/artificial-intelligence-fingerprints-security

2 Wired, "The World's First Computer Password? It Was Useless Too", Accessed 2022-01-24. https://www.wired.com/2012/01/computer-password

3 Verizon, "2021 Data Breach Investigations Report", Accessed 2022-01-24. https://www.verizon.com/business/resources/reports/dbir

4 Yubico, "Yubico Research Reveals that Cybersecurity Best Practices, Including Password Protection, and Employee Training in the UK, France, and Germany are Lackluster with the Proliferation of Employees Working from Home", Accessed 2022-01-24. https://www.yubico.com/blog/yubico-research-reveals-that-cybersecurity-best-practices-including-password-protection-and-employee-training-in-the-uk-france-and-germany-are-lackluster-with-the-proliferation-of-employees-workin

5 Transmitsecurity, "The impact of passwords on your business", Accessed 2022-01-24. https://www.transmitsecurity.com/wp-content/uploads/transmit-security-passwordless-report-the-impact-of-passwords-on-your-business.pdf
6 Gartner, "Gartner Says More Than Half of Enterprise IT Spending in Key Market Segments Will Shift to the Cloud by 2025", Accessed 2022-04-24. https://www.gartner.com/en/newsroom/press-releases/2022-02-09-gartner-says-more-than-half-of-enterprise-it-spending

4

Cyber Threats

Introduction

Cybercrime is the greatest threat to every organization in the world, and in the coming years, it is expected to become the greatest threat to every individual on the planet. As the number of people accessing the digital world increases every day, cybercrime becomes the fastest-growing criminal activity across the world.

Cybercrime is an umbrella that describes any criminal act carried out online or by using a computing device – or other electronic communication networks and information systems – as a facilitator (weapon) or as a target of criminal activity.

A cyberattack is any offensive action that targets computerized systems, IT infrastructure, and computer networks or individuals' computing devices; the aim of a cyberattack is to alter, destroy, steal data, or prevent access to information systems for malicious purposes.

Regardless of its size and industry type, no organization is immune to the growing cost of cybercrime. For enterprises, the stakes are high; according to the IBM report "Cost of a Data Breach Report 2021",[1] the cost of a data breach in 2021 is US$4.24 million. This cost is expected to intensify even further as the digital acceleration continues its rapid growth.

To have a holistic view of the current state of cybercrime activities around the world, we've compiled the following statistics to demonstrate the seriousness of this issue worldwide.

A. *Cybersecurity Ventures*[2] predicts that there will be more than 7.5 billion Internet users by 2030 (90% of the projected world population of 8.5 billion, 6 years of age and older).

B. *Cybersecurity Ventures* predicted that cybercrime damages would cost the world $10.5 trillion annually by 2025.[3]

C. Gartner predicates that 30% of critical infrastructure of organizations will be a target of cyber-attack that will result in disrupting the operation of a mission-critical cyber-physical system by 2025.[4]

DOI: 10.1201/9781003008279-4

The internet is not a safe place, and the above figures give a clear clue about the growing number of cybercrime activities and their projected damage costs in the near future.

This chapter is dedicated to covering the various types of cyberattacks; we will talk about cybercrime categories, cyberattack types, and discuss the different attack vectors employed by threat actors to access or penetrate IT systems. However, before we start, we should talk about the different threat actors commonly involved in carrying out such illegal actions and understand their motivations.

Cyber Threat Actors

The primary role of any cybersecurity professional is to defend their organization against attacks coming from cyberspace. To play this role well, they should know their adversary and understand what motivates them. This knowledge is necessary to counter their attack tactics and even mitigate them before they happen.

In Cybersecurity, a threat actor is defined as any person, group, or entity (e.g., organization, national security intelligence) trying to infiltrate the IT environment of an organization or attack individual computing devices.

For instance, threat actors can be broadly classified into internal and external actors; in Cybersecurity, we are mostly concerned about external attacks coming from cyberspace. However, some threat actors take benefit of information leaked from internal actors to execute their external attacks.

Note! A threat actor does not always have to be a person or a group of people; for example, a human error behind the misconfiguration of a server or a cloud service can cause a breach. This security incident is considered a threat actor in a broad sense; however, it is caused by non-malicious purposes.

Using the broad definition of threat actors, we can categorize them into four main groups:

Cyber Criminals

Commercial gain is the primary motivation behind this group. Cybercriminal infiltrates IT systems with malware to gain remote access to steal data (PII, PHI, financial information, and corporate data) and use stolen data later to execute more attacks or to sell on the underground markets. Another popular form of attack carried out by this group is using different attack techniques to prevent access to individuals and corporate data through encryption, and then demand a ransom to remove the restriction, as with the case of ransomware attacks.

Cybercriminals are the most significant segment of malicious actors in cyberspace; we can differentiate between individual and organized crime. Organized criminals are more organized and have the resources (money, equipment, and expertise) to launch sophisticated attacks against high-value targets, such as banks, hospitals, and public organizations. Advanced Persistent Threat (APT) and ransomware are clear examples of attacks – mostly – carried out by organized crime.

Keep in mind that some forms of hacking, such as white hackers are not cybercriminals, because they infiltrate IT systems to discover security vulnerabilities; however, cybercriminals attack IT systems for malicious intents.

Hacktivists

The term hacktivist is composed of two words, hacker and activist. Hacktivists are motivated by a social, ideological, environmental, or political cause more than monetary gain and use tactics employed by cybercriminals – although they are not very technically sophisticated – to enforce their agenda.

While traditional criminals care about stealing personal and corporate data for malicious purposes, hacktivists are more oriented towards vandalism acts; they usually launch DDoS attacks to bring target (e.g., political figures, celebrities, government's entities, business corporations, religious groups, and child pornography) website down to draw the public attention to their case.

Hacktivist works as individuals or in groups; they usually have the required funds to plan and execute their attacks. No one should underestimate hacktivists' ability to bring damage to IT systems, as they can bring severe damage – both commercial and reputation – to their targets. Some popular hacktivist groups: Anonymous and Chaos Computer Club.

Insider Threats

Insider threat is any security risk originating from within the target organization; an insider can be an employee, former employee, business partner, contractor, consultant, service provider, board member, and any party having a relationship with the target that allows them to have legitimate access to its IT systems and data. There are different motivations for insiders to reveal sensitive information about their enterprises; the reason might be for revenge, for monetary gain (selling stolen data to competitors), or only as a mistake by not following the enforced IT security policies and best security practices. Insiders pose a significant risk against target organizations, as they have information about their IT infrastructure, types of security controls, and where the sensitive data is stored, and what vulnerabilities are still exposed.

Detecting malicious insiders is a challenging task for any organization, as they can acquire information during their regular work practice without drawing attention. The best countermeasure against insider threat is to implement the Principle of Least Privilege (PoLP) in your organization; by giving each employee only the required access permission according to their job duties, and ensuring to cancel former employees' accounts as soon as they leave the organization.

Nation and State Actors

Nation-state actors have the required resources and expertise to launch sophisticated attacks against any target online; a state-sponsored attack includes any cyberattack conducted directly or indirectly (receive direction, funding, or technical help) with backing from a foreign government; such attacks commonly target big enterprises and research centers with the aim of stealing sensitive information.

Nation-state actors employ advanced stealth attack techniques to maintain their persistence in the target network for a long time; the ultimate purpose is to collect as much information as possible about the target for an extended period of time.

Most nation-state attacks are not motivated by commercial gain; these attacks are primarily conducted to steal trade secrets, military intelligence, political espionage, and even to interfere in the political process of another country (e.g., the claims of Russia interference in the USA election). Other attacks aim to destroy target data to cease its operation and cause extensive financial damage (e.g., *NotPetya* is a destructive disk wiper that works by encrypting target files and then destroying the decryption key; security experts consider it a kind of cyber weapon, not ransomware[5]).

APT is an example of a cyberattack that is commonly linked to nation-state threat actors; it will be covered in detail in the next chapter.

Terrorist Organizations

This threat actor uses cyberspace similar to organized crime and for the same purposes; however, terrorists exploit the internet in more ways such as for communications, recruiting new fighters/members, spreading their propaganda, to gain intelligence from public sources (e.g., using Open Source Intelligence (OSINT)), and buying/selling arms, drugs, false governments documents and conduct all kinds of illegal activities in the dark web.

Script Kiddies

Script kiddies are individuals attacking information systems for no clear motivation; they simply want to show their ability to infiltrate IT systems and personal computing devices. People who belong to this category do not possess advanced technical skills; they use hacking tools developed by others and execute them against target systems. Although most script kiddies' attacks are just conducted for experimental purposes, some attacks may cause a data breach, costing the target organization money and reputation damage.

In part three of this book, we will list various countermeasures (for both endpoints and enterprises) to protect against attacks raised by each threat actor already mentioned.

Categories of Cybercrime

Cybercrimes can be categorized based on who gets affected by digital crime.

Crimes Against Individual

This type of cybercrime affects any person or their properties, such as stealing personal and banking information and misusing it to harm a victim, spreading child pornography, email harassment, cyberstalking, social engineering (SE), phishing, and human trafficking.

Crimes Against Organizations

This is the most common type of cybercrime; the online presence of the target organization and its IT systems and servers become the target of the cyberattack. Examples of crime against organizations include distributed denial-of-service (DDoS) attacks, data breaches, ransomware, stealing sensitive data and selling it to other competitors or on the darknet, and copyright infringement by distributing pirated products and/or software of the target company.

Crimes Against Government

This is the most severe crime; it is also called cyber terrorism. In this type, the threat actor attacks public entities (government utilities like electricity grid, telephone and mobile communication networks, broadband internet services, water companies), and tries to exploit and shut down government websites, and utilizes the internet to spread propaganda and to fund terror activities to harm people without discrimination.

Cybercrime Attack Types

As we already said, a cyberattack involves using computer systems and networks to commit an offense against target computer systems, networks, IT infrastructure, and personal computing devices. There are different types of attacks carried out in that way. This section will list the most popular attacks that fall under this definition.

> **Note!** Many cyberattack types are also referred to as attack vectors. An attack vector is a method utilized by black hat hackers to gain unauthorized access to a computer or network server for malicious intents. For example, phishing emails are considered a cybercrime; however, it also used as an attack vector to infect with ransomware or other malware types. This section will mention all cybercrime attack types and relevant attack vectors used by cybercriminals.

Hacking

Hackers exploit weaknesses in computer systems and networks to gain access to victim computing devices (computers, smartphones, tablets, IoT devices) or networks, whether the victim is an organization or an individual. Hackers infiltrate systems to acquire sensitive corporate or personal data (PII, PHI, and financial information) for malicious purposes or aim to shut down the target website/online service – as in the case of Denial of Service Attack (DoS). Hackers have various hacking tools at their disposal to use against their targets, and they commonly possess advanced networking and security skills.

Although the term hacking is usually associated with people committing criminal acts online, the bad guys are generally distinguished by naming them Black Hat hackers.

Black hat hackers employ various technical techniques to infiltrate systems, the most common ones are: Viruses, Trojans, worms, rootkits, botnets, ransomware, browser hijacks & DDoS.

Computer Virus

A computer virus is what most people refer to when talking about malicious software. Malware, or "malicious software", is any computer program designed to damage or steal data from computer systems. Computer viruses can replicate and spread via email attachments, internet file downloads, and malicious links from one device to another. To infect a device, a virus needs a host file (program, file, or document) to carry it; the virus launches after a user executes the host file (if it is an executable program) or opens it (if the host file is a document such as MS Office or PDF).

Old computer viruses were designed to harm infected devices by altering some system functions (changing desktop wallpaper, displaying pop-up windows, disabling internet connection, changing home page, etc.). However, as businesses become more reliant on the internet and computerized systems to do their job, criminals shift their attention to creating malicious software for financial gain, such as stealing malware and ransomware.

Worms

Computer worms are malicious programs that replicate themselves from one device to another; the difference between worms and viruses is that worms can replicate themselves and spread without human intervention. Besides, the worm does not need to attach itself to a legitimate program in order to run.

Worms usually infect systems via email attachments; however, there are other methods to infect, such as malicious links sent via internet messaging applications. Once executed by the user, a worm will silently infect the system, begin replicating itself on the target system, and move to shared computers across the infected network.

Simple worms replicate themselves and aim to consume target computer resources like processing power, storage space, and bandwidth (when spreading across internal networks). Some worm types can cause damage by deleting data, stealing sensitive information, installing additional malware, and granting its operators remote control over the infected system.

Trojans

A Trojan is a type of malware that conceals itself in a legitimate program and executes when an unsuspecting user opens it. Trojans are commonly used by malicious actors to gain unauthorized access to the target computer, spy on it, and steal stored data. Unlike computer viruses and worms, a Trojan does not have the ability to replicate.

We can classify Trojans according to the type of harm they bring to the target device into the following main categories:

- Backdoor: Give remote control for the malicious actors over the infected machine.
- Rootkit: This type prevents regular detection programs like antivirus from capturing the malicious code or program. Some rootkit types operate at the kernel level, making detecting it very difficult by regular antivirus programs.
- Downloader: This Trojan type installs a new version of the malicious program on the infected machine.

Trojans infect systems using various ways. Commonly, it comes disguised with free internet programs, games, and screen savers downloaded from untrustworthy sites and pirated software downloaded from torrent websites. For instance, when downloading pirated software, you will find a file called PATCH or CRACK, which is used to illegally activate the legitimate software making it function as the paid version. Most activation programs come associated with hidden Trojans that will install silently once executed by the user.

Key Loggers

This is a form of malware that securely records and monitors everything a victim types on their keyboard and sends it to its operators. Some keylogger types can record the victim screen, audio via a microphone, and videos via a computer webcam.

Keyloggers come in two forms: Software and hardware. A software keylogger is just like any computer program; however, it usually comes with stealth features (it doesn't show up in either the Add/Remove Programs or the Taskbar). A Hardware keylogger is a small device similar to a USB key (see keelog.com) that works by attaching it to a victim computer, commonly via a USB port. Using this type requires physical access to the victim's machine. Nevertheless, captured log reports are sent via email, FTP, local area network; the more advanced one has Wi-Fi capability to control the device remotely and to retrieve the logged data.

Botnets

Botnets are collections of compromised – hijacked – computing devices (also known as zombie computers) controlled by a botmaster; these botnets work collectively to perform malicious tasks directed by their operator (see Figure 4.1).

Once a computer gets infected with the malicious bot code, it will join the infected network of botnets; this happens without user knowledge. The botnet master can now remotely control infected devices to launch malicious attacks (e.g., DDoS, sending SPAM, spreading ransomware campaigns, generating fake internet traffic on some websites for commercial gain) against any online target. A botnet network can contain thousands and even millions of hacked computers and Internet of Things (IoT) devices. Any computing device connected to the internet that has security vulnerabilities is subject to infection and becomes a bot.

DDoS Attack

This cyberattack aims to disrupt the regular traffic of a website, server, or online service by overwhelming it with false internet traffic more than the

Botnets Cyberattack model

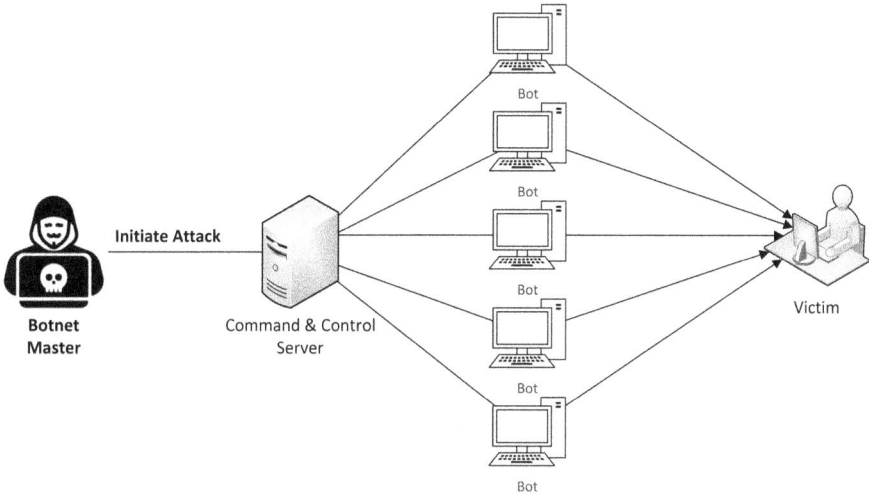

FIGURE 4.1
Malicious botnets attack model.

server, website, or online apps can handle to prevent legitimate users from accessing it. DDoS utilize botnets (see Figure 4.2) as a source of internet traffic to launch their attacks.

Cybercriminals wishing to launch DDoS attacks should not worry about infecting a large number of vulnerable devices to create botnets. Many

DDoS Attack

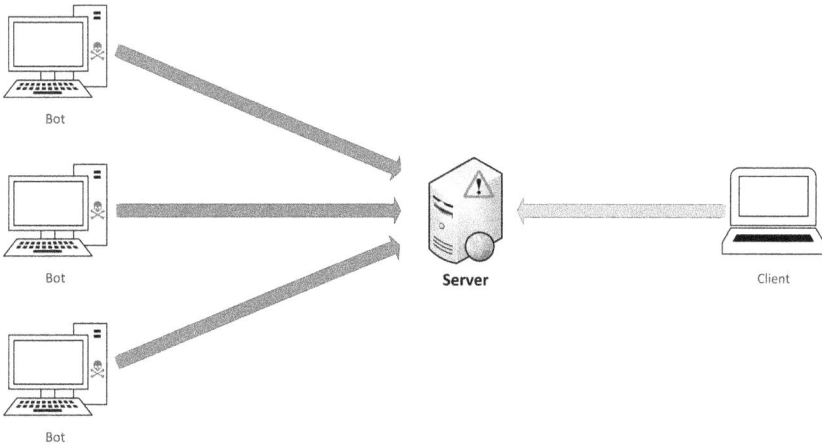

FIGURE 4.2
How a DDoS attack works.

criminal marketplaces in the darknet offer ready botnets composed of thousands of hijacked computers for sale for a few hundred dollars.

Note! According to a press release published by the Dutch Police in April 2020, Dutch authorities shut down 15 DDoS-for-hire services in one week.[6]

Ransomware

Ransomware is a type of malware that belongs to the digital extortion category of cybercrime. It works by denying access to the infected computer files through encryption; it then demands a ransom payment to remove the restriction. Ransomware is an ever-growing problem that hits all user segments like individuals, public organizations, and private businesses. Lately, ransomware attacks have become very intense, and many successful attacks have resulted in millions of USD losses to the victims. According to Cybersecurity Ventures,[7] the global ransomware damage costs are predicted to exceed $265 billion by 2031. We will cover the ransomware attack in a dedicated chapter later on.

SQL Injection

SQL injection is a web attack that works by inserting SQL queries into web form fields to control the underlying database and gain unauthorized access to sensitive information like user passwords, PII, and credit card numbers. A successful SQL injection attack can give attackers broad access to the target database to view, update, and delete stored data and to execute administrative commands against the database, such as shutting down or restarting the DBMS and even sending commands to the underlying server operating system. Although this type of attack can be – relatively – easily mitigated, many organizations are still vulnerable because they do not structure their SQL queries securely and use old programming languages to build their application web interfaces like PHP and ASP.

Man in the Middle

A man-in-the-middle attack (MITM) (see Figure 4.3) is an attack where an intruder positions himself between a conversation happening between two parties (e.g., a user and an online application). The perpetrator will relay messages between the communicating parties, making them think they are talking with each other privately, while the attacker entirely controls the conversation. This attack aims to intercept active communications to steal sensitive data like login credentials, PII, PHI, and banking information. An example of such an attack is when attackers exploit vulnerabilities in a Wi-Fi access point and deploy hacking tools to intercept all connections flowing through the access point.

Man in the middle attack

Original Connection

User

Web Application

Hacker Man In the Middle

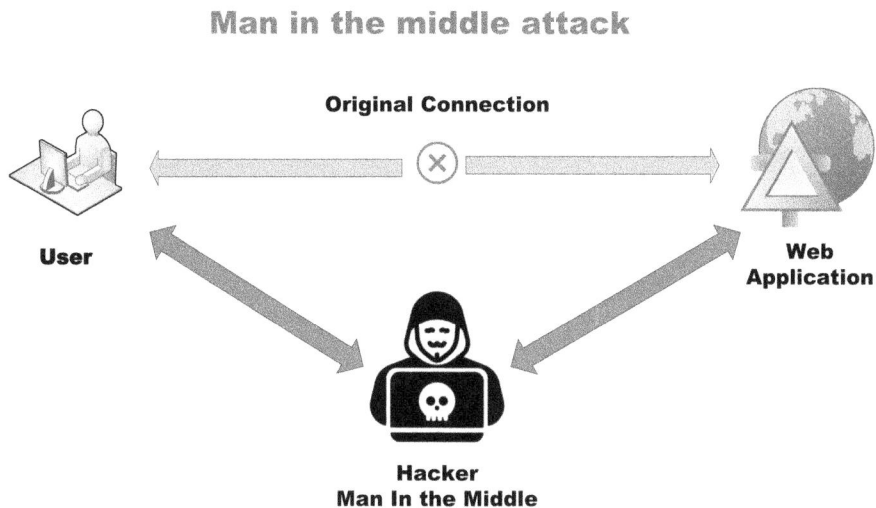

FIGURE 4.3
Man in the middle attack example.

Exploit Kits

Exploit kit is a method to detect and exploit internet users' security vulnerabilities when browsing online, automatically, and without user knowledge.

An exploit kit is a web framework sometimes hosted on compromised websites. When a victim visits the malicious site, the exploit kit will automatically scan the victim browser and operating system for security vulnerabilities (e.g., unpatched OS or outdated browser add-ons such as Adobe Flash, Java, and Microsoft Silverlight), and matching them against the repository of ready exploits stored within the exploit kit database.

Exploit kits are generally composed of the following parts:

1. *Landing page*: This is the compromised website hosting the exploit kit; here, cybercriminals determine specific criteria before moving forward with the attack. For example, check the victim's IP address to determine its geographical location; the attackers may want to target visitors from the USA only. If the victim is connecting from Russia, then no need to continue the attack. Now, suppose the victim device meets the elementary requirements of the attackers. In that case, the landing page will execute code on the victim's browser to check for vulnerable add-ons, unpatched OS, or outdated applications. If there is any vulnerability, the victim will be redirected to the exploit page.

2. *Exploit*: The exploit will use the vulnerable application to execute the malware on the victim device.

3. *Payload*: if the exploit was successful, the exploit kit would send a payload to infect the victim device. The payload can be the malware itself or a downloader that will download another malware. An advanced exploit kit delivers the payload encrypted to prevent network monitoring systems from detecting it and then decrypts the payload on the victim's computer. The most common payload delivered via an exploit kit is ransomware.

Cybercriminals are now offering exploit kits for rent (this criminal model is known as Exploits as a Service) in some darknet marketplaces. It allows affiliate criminals with average technical experience in hacking to use such kits – which come with an easy-to-use management console – to spread malware and infect victims.

The most famous exploit kits are Angler, Neutrino, Magnitude, Rig, and Nuclear Exploit Kits.

Drive-by-Downloads

A drive-by download is a download that occurs without user knowledge. Threat actors utilize drive-by-download by placing the malicious code on their website or by infecting another –legitimate – website with it by exploiting some vulnerabilities. When the unsuspecting visitor accesses the compromised website, it will scan their device for known vulnerabilities and automatically execute the background malware. Ransomware operators commonly utilize this attack technique to infect their targets.

Malvertising

Also known as "Malicious Advertising" it is the act of using internet advertisements to deliver malware to end-user devices. Malvertising works by injecting malicious JavaScript code inside legitimate online ads. Now when a visitor loads the page (or a spam email) with the malicious ad or click on it, it will redirect their browser to a phishing page or to a page hosting an exploit kit that will take advantage of any security vulnerability existing in the visitor web browser or its operating system to infect with malware.

Combating Malvertising is difficult, as perpetrators use legitimate online ad networks such as Google AdSense to deliver their malicious ads. Many of these ads appear on highly-reputable websites such as news portals and online merchant sites making visitors trust to click on them. Malvertising can also infect victim computing devices without requiring any interaction from the visitor side (such as clicking on links).

The complexity of online advertisement makes spreading Malvertising possible. For instance, many large websites outsource their advertisement space to outside vendors, who will, in return, resell it again to other third-party vendors. Some of these vendors have developed self-publishing platforms that allow individuals to buy ad space; criminals can abuse this service to place malicious ads on various trustworthy sites.

Malvertising can introduce various malware types into victim devices, such as ransomware, spyware, adware, computer viruses, and malicious crypto mining (Cryptojacking).

Cryptojacking

Also known as "Malicious Cryptomining" this is a modern form of malware that aims to steal victim computing devices' resources (CPU and RAM) to mine cryptocurrencies (e.g., Bitcoin) for commercial gain.

> Note! Cryptocurrencies are virtual money that only exists in the online world without any physical presence. The most well-known cryptocurrency is Bitcoin, which first appeared in 2009.

Cryptojacking malware is designed to be fully hidden from the user; it works in the background and consumes computing power to solve tricky cryptographic puzzles that finally result in profiting its operator with virtual coins.

Understanding how criminals use the computing power of unsuspecting victims to profit from mining cryptocurrency requires a detailed explanation of how cryptocurrencies work in general and how the crypto mining process works outside the book's scope. So, we advise readers wanting to expand their knowledge about the subject to find relevant books on the CRC website.

Identity Theft

Identity theft is the worst attack against individuals online. In this type, a criminal steals your data such as your full name, credit card number, email address, driving license, and government ID to impersonate you to commit fraud, buy goods and services online using your personal info, and conduct all types of illegal activities using your name.

There are different types of identity theft:

1. Financial Identity Theft: This is the most common identity theft; attackers impersonate the victim to get a credit card or bank loan using their name.
2. Medical Identity Theft: Some criminals steal the victim's personal information to get healthcare services such as medical care and drug prescriptions.

3. Criminal Identity Theft: In this type, a criminal handles false government ID using the stolen personal information of someone else to direct criminal charges to the victim instead of the real criminal.

4. Child Identity Theft: This crime happens when someone steals minor personal information (especially the Social Security number) and uses it to commit different kinds of work, such as issuing a driving license or taking a loan from a bank.

5. Identity Cloning: In this type, the identity thief tries to impersonate someone else to hide for any reason. For example, some people use someone else's photo on their social media accounts to conceal their accurate picture. This is a kind of identity cloning that is difficult to discover.

Identity theft is a significant problem for individuals and becomes more problematic with the rise of the information age and the widespread adoption of technology in all life aspects.

Scamming

Scamming is an attempt by criminals to sell faulty products/services online to victims. Scammers use email messages, social media platforms, and sometimes direct phone calls to convince a victim to buy something false or try to acquire personal information from victims to use it for other malicious purposes. For example, scammers use different tactics to commit fraud against their victims, such as:

1. Offer online IT support services.
2. Offer work-from-home opportunities with a large salary.
3. Offer travel vacations at a low cost.
4. Offer high valued products with low prices on fake shopping websites.

SPAM Email

SPAM email is also known as Junk email; it is the act of sending unsolicited email messages in bulk to a large number of recipients without their permission. Spammers deceive users by offering various promotions of various products and services to convince them to visit their site or respond to their email to get more information.

Most SPAM emails are sent for commercial purposes. However, a portion contains infected attachments or links to malicious sites hosting malware or exploit kits to infect visitors with different malware types (e.g., ransomware, spyware).

SPAM is considered illegal in all jurisdictions worldwide; however, there is no 100% foolproof solution to stop it until now.

Social Engineering

Social engineering is a term used to name various malicious attacks accomplished through human interactions. SE uses psychological tricks to manipulate victims' minds and convince them to reveal sensitive information to the attacker.

Attackers need first to collect some background information about the intended target to use it as an entry point to initiate the attack. The internet gives a wealth of information about any target online. For instance, attackers can utilize different OSINT techniques to collect information about individuals from online public sources, such as social media profiles, public databases, and internet archive records.

There are different SE attacks, depending on the medium used to conduct the attack; for instance, attackers use email messages, social media, phone calls, and even USB drives to begin their attacks. SE attacks and gathering intelligence from public sources will be covered in a dedicated chapter later.

Phishing

Phishing is the most well-known type of SE attack, usually conducted via email messages, social media, internet messaging, and even SMS messaging. Phishing attacks aim to steal sensitive information such as account credentials, social security numbers, credit card details, and other valuable personal information to get unauthorized access or use for malicious purposes such as fraud and identity theft. Some phishing attacks do not stop by stealing victims' personal information; they also infect their devices with viruses and malware.

In a typical email phishing scenario, intruders masquerade as a trusted entity such as your bank, social networking site, internet service provider, online payment website, or company you have business with. Then ask you to update your account information details by clicking on a link provided in the email. Now, if the unaware user clicks on that link, it will take them to the completely identical phishing website (complete clone) of the original one. When a victim enters their credentials in the login form, attackers record this information and redirect the user to the original website again.

> Note! Phishing is a massive problem for internet users and organizations alike, especially after the wide adoption of the remote-working model. According to KnowBe4, 91% of cyberattacks begin with a spear-phishing email.[8]

There are different types of phishing, such as Spear Phishing, Spear Phishing Attachment, and Whaling. In a later chapter, we will talk about phishing attack types and how we can spot them.

Cyberstalking

Cyberstalking is the act of using the internet (email, forums, social media messages, internet messaging applications) or another electronic medium to stalk or harass another person, group, or company. Stalker uses different tactics such as false accusations, defamation, slander, and threats to control the victim's mind or force them to reveal sensitive information that can be used later to make offline stalking or identity theft.

What makes cyberstalking relatively easy is widespread social networking sites; users of these platforms post extensive personal details about their personal life and family, including private photos and videos. On the other hand, the advance in communication technology has simplified our lives, making it possible for us to chat with another person – via different internet messaging Apps (e.g., WhatsApp) – located in another state or country from the comfort of our bedroom. The amount of personal information available publicly and the ease of making communications online are what a stalker needs to begin their malicious action.

Cyberstalkers use online anonymity techniques (e.g., VPN and TOR Browser) to conceal their digital identity. Many cyberstalking cases involve a criminal impersonating an actual person to fool the victim before conducting the stalking act.

Cyberbullying

Cyberbullying is the act of using the internet to bully someone else. Cyberbullying occurs online on social media sites, discussion forums, online gaming, emails, IM Apps, or via cell phones such as SMS and direct phone calls. Cyberbullying involves posting harmful content online (sometimes including personal details about the victim) that hurts someone else; the bullies aim to damage the victim psychologically by causing embarrassment or humiliation.

Similar to cyberstalking, this crime is very prevalent online for the same reasons (prevalence of social media sites and the ease of reaching people online anonymously). Cyberbullying is considered an offense in most countries.

Software Piracy

Software piracy is the illegal act of copying software programs and using them in a way that violates their license agreement. Software piracy can be branched into the following types:

1. Counterfeiting: This is the act of cloning and distributing software programs without the owner's permission. For example, many users in third-world countries copy a Windows OS disc into hundreds of discs and sell them for a low price.

2. Internet Piracy: Similar to counterfeiting, however, this type refers to programs downloaded from the internet. For example, peer-to-peer networks such as torrent websites offer a wide selection of pirated content such as software programs, eBooks, and movies available for free download.

3. End-User Piracy: This happens when a legitimate user violates the end-user agreement of the purchased software. For example, many programs are licensed to be installed on one device only; if a user installs them on two or more devices, this is considered a violation of the program license agreement.

4. Client-Server Overuse: Some software programs are designed to work on a network by installing it on a central server and licensed for a specific number of users. For example, suppose a database management system is licensed to allow 100 employees to access it concurrently, and the company increases the number to 120. In that case, this is considered a violation of the license agreement.

5. Hard-Disk Loading: Some technology stores install pirated copies of programs and operating systems on their computers to make them more attractive for them to buy.

Although it is not considered a type of software piracy, duplicating eBooks, magazines, and copyrighted content and selling/distributing it online is regarded as a crime. This crime is known as Copyright infringement.

Cyber Attack Vectors

The previous section covered the different types of cybercrime attack types that criminals conduct. This section will list the primary attack vectors (or delivery methods) that criminals utilize to attack and infect other people with malware or steal their money and sensitive data.

Note! In cybersecurity, we should differentiate between attack vector and attack surface.

Attack vectors: These are the methods cybercriminals employ to attack and compromise computer systems.

Attack surface: This is the number of available attack vectors that criminals can exploit to target networks and computer systems.

Some attack vectors carry the same name as the cybercrime conducted using them; for instance, we have already covered the following cybercrimes, which are also mentioned in the context of attack vectors:

1. Malicious emails
2. Malvertising
3. Malicious software as a service
4. Pirated software
5. Botnets
6. Exploit kits

In addition to the previously mentioned, the following attack vectors are utilized by cybercriminals to conduct their attacks.

Note! In general, malware can access the target system using one of the following two ways:

   1. Exploit a vulnerability in the victim computing device
   2. Deceiving a victim to execute the malicious program.

Removable media

Although it belongs to the old hacking school, attacking computer systems via USB and other removable media devices is still used actively by cyber-criminals. For instance, USB malware has been developed to infect devices silently with a limited possibility of being detected by typical antivirus solutions. Besides, some malware types like Worms, ransomware, and viruses can propagate across local networks. For example, when an unaware user inserts a compromised USB key into one computer, the infection will spread to all connected devices across the network. USB infection is considered a local – or insider – threat. It requires physical access to the target device; however, criminals invent different ways to fool victims into inserting malicious USB keys into office and home computing devices.

Note! A famous cybersecurity attack conducted using USB devices happened in 2010. The *Stuxnet* worm has used a USB device to spread malware into the network of an Iranian nuclear facility.

Office Macros

The Office macros feature is a way to extend MS Office functions by adding programmatic instructions using Visual Basic for Application (VBA) programming language. This allows MS Office users to perform many added functions and automate frequently used tasks to increase productivity.

Cybercriminals exploited office macros to infect malware many years ago; for instance, perpetrators used to send Microsoft Office files with macros via email attachments or post them to online forums. These files were named to make victims trust them (e.g., invoice, receipt, legal agreement, and terms). Once the unaware user opens any of these files, the macro will run automatically and download the malware into a victim device.

In modern versions of Microsoft Office, macros are disabled by default. However, cybercriminals are creative in employing different SE tactics to convince unsuspecting users to enable macros in received files and, consequently, infecting with malware. A typical tactic works by displaying a fake warning upon opening the file asking the user to enable the macros to read the file correctly. Till now, Office macros are still the preferred attack vector of most ransomware families.

Zero-day Vulnerability

A Zero-day (also known as 0-day) vulnerability is a newly discovered software flaw – or bug – which is still undiscovered by the software/antivirus vendor. Cybercriminals can exploit such flaws to gain unauthorized access or to spread malware.

Zero-day vulnerabilities are flaws discovered in software applications (e.g., PDF readers, web browsers, compression programs) and operating systems; these security holes can result from programming errors, misconfigurations of software programs, or improper operating-system configurations. If a software vulnerability has not been addressed on time, hackers could exploit it to conduct malicious acts; this attack is a zero-day attack.

Zero-day vulnerabilities are rare; you cannot expect to see one every day. Besides, they are not only important to hackers and organized crimes, but nation-state actors are always interested in exploiting this type of vulnerability.

Many organizations offer bounties for capturing zero-day vulnerabilities in their software/systems; they provide attractive cash rewards for anyone to report such vulnerabilities directly to the vendors.

Note! Sometimes, cybercriminals offer zero-day vulnerabilities for sale in the darknet marketplace.

Lack of Cybersecurity Training

This is not a technical attack vector. However, it is considered the most critical element (the human element) in any cybersecurity defense plan and is the direct cause of most cyberattacks targeting individuals and organizations' IT systems and networks.

A survey conducted by *GetApp*, found that only 27 percent of companies provide SE awareness training for their employees. The same study also found that 43% of employees do not get regular data security training, while 8% reported never receiving any training.[9]

Any organization that operates in today's information age should have a cybersecurity awareness training program covering all subjects that cybercriminals can exploit, such as phishing, mobile security, physical security, password security, and others. This helps employees understand the different cyber threats they may face at work and home and the best methods to mitigate them. Training empowers employees with the required knowledge to remain up to date with the latest cyber threats and deploy the best security practices and defense techniques. Besides, training helps companies enforce their security policy rules on their employees by making them aware of the importance of implementing such policies to minimize business risks and comply with various regulatory compliance standards.

Many companies only direct their cybersecurity awareness training to the IT team while leaving other managerial roles (such as sales and marketing) outside this endeavor. However, this doesn't seem right, as any employee who has access to company IT systems should have adequate training that matches their job duty – or role – in the company.

Some companies prefer to outsource their cybersecurity awareness training to a third-party provider. In contrast, big organizations with adequate resources and budgets – prefer to develop their in-house training programs.

Summary

This chapter was dedicated to covering the various types of cyberattacks; we began talking about the cyberthreat actors and what motivates each one to conduct their malicious activities. We categorized cybercrime – according to those affected by the digital crime – into crimes against the government, individual, or enterprise.

There are various cybercrime attack types; we mentioned 20 of them and discussed how each one works. Attack vectors are the methods employed by cybercriminals to attack computer systems and networks. We listed the different attack vectors used by criminals to launch their attacks and emphasized the importance of having a cybersecurity awareness training program for each organization to teach employees how to defend themselves and the organization against threats.

Next, we will dedicate the following three chapters to discussing three critical cyberattacks: Ransomware, APT, and SE attacks. Ransomware and APT attacks have been very active lately; they cause millions of losses to organizations worldwide. SE attacks target both individuals and organizations and are considered the main entry point of most devastating attacks that target IT systems and networks.

Notes

1 IBM, "How Much Does a Data Breach Cost?", Accessed 2025-04-02. https://www.ibm.com/security/data-breach
2 Cybersecurityventures, "Cybersecurity Ventures Official Annual Cybercrime Report", Accessed 2025-04-02. https://cybersecurityventures.com/hackerpocalypse-cybercrime-report-2016
3 Cybersecurityventures, "Cybersecurity Ventures Official Annual Cybercrime Report", Accessed 2025-04-02. https://cybersecurityventures.com/hackerpocalypse-cybercrime-report-2016/
4 Gartner, "Gartner Predicts 30% of Critical Infrastructure Organizations Will Experience a Security Breach by 2025", Accessed 2025-04-02. https://www.gartner.com/en/newsroom/press-releases/2021-12-2-gartner-predicts-30–of-critical-infrastructure-organi
5 Knowbe4, "[ALERT] NotPetya is a Cyber Weapon, Not Ransomware", Accessed 2025-04-02. https://blog.knowbe4.com/notpetya-is-a-cyber-weapon-not-ransomware
6 Politie, "Politie houdt verdachte aan voor DDoS-aanval op MijnOverheid.nl", Accessed 2020-04-15. https://www.politie.nl/nieuws/2020/april/10/03-politie-houdt-verdachte-aan-voor-ddos-aanval-op-mijnoverheid.nl.html
7 Cybersecurity Ventures "Global Ransomware Damage Costs Predicted to Exceed $265 Billion by 2031", Accessed 2022-04-24. https://cybersecurityventures.com/global-ransomware-damage-costs-predicted-to-reach-250-billion-usd-by-2031
8 Knowbe4, "91% of Cyberattacks Begin with Spear Phishing Email", Accessed 2025-04-02. https://blog.knowbe4.com/bid/252429/91-of-cyberattacks-begin-with-spear-phishing-email
9 Getapp, "Social Engineering Techniques that Hack Your Employees", Accessed 2025-04-02. https://www.getapp.com/resources/social-engineering

5

Advanced Persistent Threat (APT)

Introduction

As the world digitalizes, internet security and the best methods to protect the IT systems of businesses and public entities become widely discussed. Despite billions of dollars spent annually on cybersecurity defenses, the threat of Advanced Persistent Threat (APT) – which stays undetected for months and even years – remains a significant concern for organizations on a global scale.

APT attack is commonly linked to a group, such as an organized criminal group, a nation-state actor, or a state-sponsored group. Such a group has the capability, resources, and intent to launch sophisticated attacks against the IT system and computer networks of valuable targets – big enterprises or government entities – to infiltrate IT environments and gather as much information as possible for an extended period without being discovered.

The motivations behind APT attacks are mainly political or economic. For example, stealing trade secrets and sensitive financial data, spying on the military/defense sector of another state, capturing diplomat messages, obtaining credentials to critical infrastructure IT systems, or disrupting (sabotaging) vital services (e.g., attacking telecommunication networks, sabotaging electricity grid and water supplies, database deletion or complete site takeover in addition to sabotaging nuclear and petrochemical facilities) of the target are all examples of APT attacks driven by such motivations. When the APT aims to damage critical IT systems, it becomes a type of Cyberwarfare act.

APT attacks are not linked to individual actors. Such attacks need a high level of coordination between many people who possess vast skills in complex hacking operations, intelligence gathering, and social engineering tactics. Besides, it is not easy for individuals to maintain stealth access (persistent) for a long time after intruding into the target IT environment. Some APT attacks require intelligence from non-internet sources, such as the information acquired from target organizations' internal sources.

An APT attack needs extensive human and financial resources to conduct than common cyberattacks; this is why most APT attacks are linked to state-sponsored actors.

DOI: 10.1201/9781003008279-5

Note! APT attacks are not limited to targeting private companies and public entities; individuals such as politicians and activists are among their targets.

This chapter will define the APT term, understand its indicators (or symptoms), list the typical phases of APT attacks, and talk about major APT groups. We will defer talking about detecting and preventing APT attacks to a later chapter.

What Is APT?

To fully understand the APT attack's precise meaning, we should understand each of the three terms that constitute the APT acronym.

1. *Advanced*: APT operators have access to vast intelligence resources from open and commercial sources. This knowledge appears clearly in how they craft their spear-phishing campaigns and choose their targets when launching social engineering attacks to gain an entry point to the target environment. An APT utilizes many tools ranging from standard attack tools to developing custom tools/exploits customized to fit the target IT system. For instance, APT operators use various security and system/network administration tools which are not initially created for hacking purposes, such as network discovery and password recovery tools. However, they utilize them to discover the target network, identify the software and hardware used, and select the best techniques to gain access. APT may use ordinary malware (procured from available malware construction kits) to remotely infect or control a system. However, APT attackers use undiscovered vulnerabilities in some attacks, referred to as zero-day exploits, and sophisticated malware explicitly built to target a specific IT infrastructure and circumvent any security measures implemented. Using custom attack tools is another clue to the progressive nature of APT attacks that distinguishes them from other ordinary cyberattacks, which tend to use general automated attack tools.

2. *Persistent*: APT operators aim to maintain extended access to the target environment. They do not pursue immediate financial gain or cause damage to the target network. Instead, they want to remain undetected to gather as much information as possible over an extended period. APT operators are persistent, and if they fail in one attempt, they will repeat the attempt till they succeed.

3. *Threat*: the APT attack is a severe cyber threat executed by humans in most of its phases. It targets large organizations and government entities to steal the most valuable data of the target. APT attackers use advanced hacking tools and exploit zero-day vulnerabilities to infiltrate the target organization's network; however, what makes an APT attack successful depends on the creativity of the attackers. APT attacks are well-funded and executed with precise coordination between highly skilled groups of attackers –instead of mainly relying on automated hacking tools and scripts (e.g., worms, bots) compared with traditional cyberattacks.

APT Indicators

Different indicators differentiate an APT attack from other traditional cyberattacks. We will list them below.

1. *Resources intensive*: Invading a well-protected target with an APT attack may require considerable money (the cost may range from thousands to millions of USD!). It includes the cost of training, scanning, and discovering vulnerabilities in the target environment, customizing the attack tools, and gathering intelligence from public and commercial sources about the target entity and its essential employees. Some APT actors purchase intelligence information from internal actors (spies) within the target organization to guarantee a high success rate for their exploitation. A single APT attack may require months of preparation before launching it.

2. *Focus on objectives*: APT attacks last for an extended period to perform only the specific task/s set by their sponsors. For example, during an APT attack, the attackers may discover some customers' personal information. However, they may not exploit this info (for commercial gain) and focus on their operator's primary goal, which could be stealing sensitive trade secrets and remaining undetected.

3. *Operators skills*: APT operators possess advanced technical and social engineering skills. They are professional in creating malware that evades antivirus and firewall detection; they know how networking appliances – from different vendors – operate and have a deep understanding of how to remain anonymous to send the stolen data silently outside the victim organization's network. The advanced level of expertise of the attackers primarily distinguishes APT attacks.

4. *Big targets*: APT attackers commonly target large entities such as research centers, universities, healthcare/pharmaceuticals, big industries (oil, energy, telecom, mining, high-tech, chemical), and government entities working in sensitive sectors.

5. *Attack timeframe*: APT attacks are known to last for months and even years, unlike common cyberattacks that stop after gaining access and performing malicious actions (e.g., data breach or infecting malware, such as ransomware).

6. *The number of attack points*: APT operators spend months discovering target environment vulnerabilities. They collect information about its IT infrastructure and type of security defenses in use and gather intelligence about essential employees (gatekeepers) who have access to critical resources in the target organization. This knowledge allows the attackers to exploit the weakest point to gain entry.

7. *The number of people involved in the attack*: As we already said, the number of attackers in APT attacks is relatively big. A harmonization between those attackers is necessary to achieve successful penetration and persistence.

8. *Method of entry*: APT attacks commonly begin with a phishing attack – as a part of a large-scale attack – to gain an entry point into the target IT environment. In such a scenario, an APT attack begins by compromising one employee's device after conducting a careful reconnaissance first and then using the infected device as a base to circumvent security defenses and gain access to the target environment to perform its malicious actions.

9. *Targeted attack*: An APT attack is targeted, while other traditional cyberattacks are not necessarily targeted.

APT Warning Signs

In the previous section, we saw what differentiates an APT attack from other regular cyber-attacks. APT attacks are challenging to detect because they use a plethora of automated attacking tools and require sophisticated human skills to conduct. In this section, we will talk about the most prominent signs of APT attacks; when discovered, they should draw your attention to the possibility of an APT attack.

Spear-phishing Emails

Spear-phishing is a type of phishing scam; it is a targeted attack and is generally directed toward individuals in the target organization. Spear-phishing email aims to steal the account credentials of key employees (e.g., high-level business executives or IT managers) who have broad access to the target organization's network or plant malware on the target computer – as with the case of Spear-phishing attachment attack.

Spear-phishing emails are crafted very well. Attackers gather information (work and personal information) about their target first and then prepare an email message accordingly. APT attackers use different social engineering tactics to convince the user to respond to the phishing email to expose sensitive information or open an attachment, such as an MS Excel file containing malware.

Spear-phishing emails remain the favorite attack technique that APT actors utilize to infiltrate target networks. After compromising the endpoint device of the target, they begin to move laterally across the IT environment to access more sensitive resources.

Strange Logins

Another symptom of APT attacks would be a strange login to the system after work hours, especially if high-level roles (IT manager, IT administrator, CEO) conducted these logins in your organization. APT attackers select a time when key employees responsible on monitoring networks are not present; many attacks are conducted on weekends and holidays. Detecting unusual signs in such unusual timing could be a worrying indicator of an APT attack.

Discovering Specific Malware Types

The presence of some malware types, such as backdoors and spyware, could be a warning sign of an ongoing APT attack. APT attackers try to maintain a long presence in the target environment even if the compromised accounts they use for initial access have changed. These malware allow them to keep extended access and open a gap to exfiltrate information and receive attack instructions.

Unusual Data Traffic

The primary motivation of APT is stealing sensitive information. If we discover unusual data movement between computers within our network, and if a large amount of data is moving outside our organization's network outside work hours, this could be an APT attack sign. Remember that APT may not shift the data instantly outside the target environment; instead, they may move the data from the storage server to a specific endpoint device and later move it outside.

Any connection to external computers should be treated as suspicious, especially when happens outside work hours and originates from ordinary computers that are not supposed to move data to external locations.

Strange File Types

APT attackers collect data and group them into large compressed files, waiting for the chance to move them outside the organization's network. If you discover large compressed files that are not supposed to exist in some places, be careful because this is a warning sign of an APT attack.

Accessing Unusual Domains

Monitor your network access logs for unusual web browsing activities. For instance, this should be a warning sign if you discover internal communications with suspicious domain names or with domain names with less popular extensions (such as design, art, biz, mobi, skin, video).

Storage Capacity Decrease

As we already mentioned, adversaries will gather data and store them in one location (such as an endpoint device) before moving them outside the organization. If you discover the storage capacity of a specific endpoint device is heavily low, while you do not see files within it (APT attackers will hide their stolen files), this should make you feel suspicious.

APT Attack Life Cycle

A typical APT attack scenario will work as follows:

1. *Select the target*: The first phase of an APT attack begins by gathering initial intelligence about the target entity (individual or organization). Attackers select their target via two methods: Using Open Source Intelligence (OSINT) techniques to gather intelligence from publicly available sources. For example, the target company website can be utilized to discover previously published information by checking some web services that record previous versions of any webpage. The revealed information may contain previous contact details of employees (phone numbers and email addresses), and metadata info of photos and videos pulled from the archival version of the target website. *Wayback Machine* (https://archive.org/web) is an example of such an archival service. APT attackers also utilize specialized tools such as *Netcraft*, *ARIN*, and *DNSstuff*, and job websites that list open vacancies in the target company to discover technology infrastructure. The second method works by selecting victims accidentally. For example, attackers conduct general online research to discover organizations that are still using vulnerable IT systems and/or other exploitable software products; once found, they begin planning their attack.

2. *Organize the attacker team*: Once the target IT systems have been identified, the second phase forms the attacking team. APT attacks require sophisticated skills, so APT operators hire skilled professionals in their work area to attack with minimal possibility of being discovered.

3. *Build and acquire attack tools*: After forming the team, attackers begin preparing the attack tools, which can be built specifically for the target technology infrastructure or procured from common cyberattack tools.

4. *Information Gathering*: Before launching the attack, APT must gather as much information as possible about the target IT infrastructure; this includes target network topology, DMZs, internal DNS and DHCP server, the technology used to build the target website and other IT systems, hosting provider, email system, VPN solution, open ports and services, security defenses that the target possesses, etc. Identifying all this info is not an easy task and will require time to complete depending on the target IT infrastructure size.

5. *Test response*: In this phase, the APT attacker tests the target organization's security defenses by looking at weak spots (entry points) in the network and exploiting them with a test attack tool. There are various ways APT attackers utilize to gain an entry point, such as email, discussion forums, social media sites, malicious advertisements (Malvertising), and exploiting VPN service vulnerabilities. APT's most famous attack technique is using phishing emails that include malicious attachments or links to malicious sites as a starting point to infect individual systems before spreading to other locations. APT usually uses personalized phishing emails tailored specifically to a specific person – who has high credentials – to increase the likelihood of opening the email and the attached file. This type of attack is called Spear phishing, and it depends on the information gathered about the target during the "Information Gathering" phase.

   Here is an example of a Spear-phishing email (see Figure 5.1).

6. *Deploying attack tools*: Once an entry point has been successfully identified and tested, the full suite of attack tools is deployed into the target network.

7. *Launch initial attack*: After entering the target network, attackers begin to discover the network and decide which target to attack first (e.g., data or email server).

8. *Set up the outbound connection (Command and Control)*: After exploiting the target, APT attackers need to establish a secure channel to offload information gathered. This channel (also known as Command & Control Server, C&C) is also used to update the malicious code of the installed malware and maintain its persistence by preventing the target's installed security solutions from detecting it.

Account De-activation Process

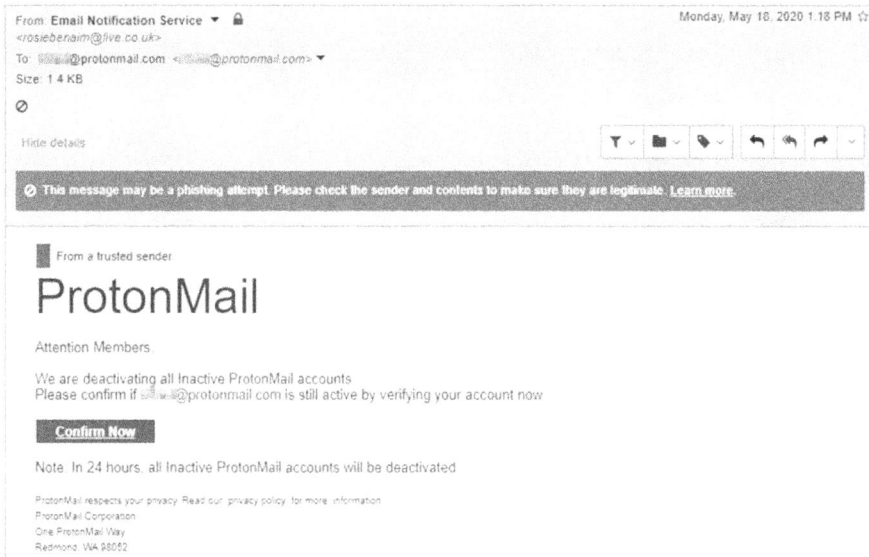

From: Email Notification Service ▼ 🔒
<rosiebenaim@live.co.uk>

To: [redacted]@protonmail.com <[redacted]@protonmail.com> ▼
Size: 1.4 KB

Hide details

⊘ This message may be a phishing attempt. Please check the sender and contents to make sure they are legitimate. Learn more.

■ From a trusted sender

ProtonMail

Attention Members

We are deactivating all Inactive ProtonMail accounts
Please confirm if [redacted]@protonmail.com is still active by verifying your account now

Confirm Now

Note: In 24 hours, all Inactive ProtonMail accounts will be deactivated

ProtonMail respects your privacy. Read our privacy policy for more information.
ProtonMail Corporation
One ProtonMail Way
Redmond, WA 98052

FIGURE 5.1
Example of a spear-phishing email trying to acquire the victim email account password.

9. *Access expansion*: In this phase, the attackers try to spread inside the target environment. To strengthen their foothold, they try to steal other users' logins and passwords (privilege escalation) for target systems to gain access into more places within the network (security and other critical systems).

10. *Exfiltration of data*: In this phase, attackers begin executing their objectives (whether stealing or destroying data), commonly by sending stolen data to their operators outside the target environment. APT utilizes encryption and other steganography techniques (hiding data within other file types) to conceal the transmission of stolen data, thus preventing victim security defenses from recognizing and detecting stolen data. Sometimes, APT attackers launch DDoS attacks during the exfiltration phase to consume victim IT resources and prevent system administrators from detecting the APT attack by drawing their attention to other areas.

11. *Anonymizing the traces and maintaining persistence*: After accomplishing their goal, attackers clear any traces that can point to their actions inside the victim network or can be used to identify the source of the attack. APT attackers may leave a backdoor to return later to continue the data exfiltration.

Famous APT Groups

The date when the first instance of an APT attack took place remains under debate, just like when the first computer virus appeared. The US Air Force originally coined the term APT in the early 2000s; however, APT groups have been in operation before this date for a relatively long time.

This section will talk about the most prominent APT groups and mention some famous attacks tied to them.

Equation Group

Equation Group is the most sophisticated APT group on earth; many security experts link it to the *United States National Security Agency* (NSA) because of this group's advanced methods and hacking tools to infiltrate target IT systems. Equation group uses a plethora of hacking tools and zero-day exploits that target major firewall and antivirus vendors. They also use sophisticated encryption and steganography techniques to hide their presence.[1] In 2015, the Russian antivirus vendor *Kaspersky Lab* documented 500 malware infections in 42 countries executed by this group; however, due to the self-destruction feature built into the malware used by this group, this number is expected to be much larger than this.

The most famous attack launched by this group is the *Stuxnet* malware. This is a computer worm spread via Microsoft Windows computers that first appeared in 2010. *Stuxnet* – which traveled via USB stick – was used to attack the industrial control systems (specifically, the programmable logic controllers (PLCs) responsible for automating machine processes) of the *Natanz* uranium enrichment facility, leading to substantial disruption of the Iranian nuclear program.

There is other malware based on *Stuxnet* source code; the following are the most known varieties:

1. *Duqu* (2011): This malware was created to steal data from industrial facilities to prepare for future attacks; it works by exploiting zero-day vulnerabilities to execute its malicious actions. *Duqu* was linked to Unit 8200 of the Israeli Intelligence. Symantec Security Corporation believed after analyzing the *Duqu* malware, that its source code is identical to the *Stuxnet* malware.

2. *Flame* (2012): Flame was an advanced spyware tool. It records everything the victim does on his/her device, including screen captures. The primary geographic area affected by the *Flame* espionage campaigns was Middle Eastern countries.

3. *Havex* (2013): Also known as *"Backdoor.Oldrea"*, this malware was linked to the Russian APT group *"Energetic Bear"*. It targets

organizations (steal information) in the following industries: energy, aviation, defense, pharmaceutical, and petrochemical sectors. Its victims were mainly based in the USA, Canada, and European Union countries.

4. *Triton* (2017): This is a murderous malware that first appeared in the Middle East and then spread to North America and other world regions. It aims to disable (or take over these systems remotely) the safety systems of petrochemical plants, preventing them from functioning correctly, which will result in causing severe injuries to workers if a problem arises during work.

Lazarus Group

Also known as *"Guardians of Peace"* or *"Who is Team"*, while the USA intelligence name it *HIDDEN COBRA*. Lazarus is a cybercriminal group and is known to be operational since 2009 when they launched a cyber-espionage campaign targeting the government of South Korea and tried to cover their traces by executing a DDoS attack. In addition to the persistent nature of their attacks, the sophisticated attack methods used in their attacks put them under the APT category. The most famous attack linked to this group is the attack against *Sony Corporation* in 2004, which exposed personal information about Sony Pictures employees and their families. This group is known to be sponsored by North Korean intelligence services.

GhostNet

This name was given to a group that conducted a large-scale espionage campaign discovered in March 2009. *GhostNet* APT group has compromised computers in more than 103 countries worldwide; they mainly target foreign embassies, diplomats, government ministries, and Tibetan institutions. Attacks were conducted primarily by sending spear-phishing emails to compromise victim devices by installing a trojan horse that controls their microphone and camera, turning victim devices into surveillance systems. The command and control servers of the *GhostNet* group were based in China; however, the Chinese government denied any relation with this group.

Note! The *Information Warfare Monitor* has published an investigative report (Tracking GhostNet: Investigating a Cyber Espionage Network) about *GhostNet*. You can find it here https://archive.f-secure.com/weblog/archives/ghostnet.pdf

Fancy Bear (APT 28)

An APT group tied to Russia. First discovered by Trend Micro in 2014; however, it is likely to have been operating since 2000. This group mainly specialized in attacking military and government targets in Ukraine and Georgia, and NATO agencies, French and German institutions, and USA defense contractors. The most notable attack of this group was its interference with Hilary Clinton's 2016 election campaign.

APT28 uses Spear phishing and *Mimikatz* (a tool for stealing Windows passwords, PINs, and Kerberos tickets from memory) to infect and access victim devices. APT28 group attacks are directed to cause political chaos; its promotion of the Russian government's political plan and its advanced attack methods classify it as a state-sponsored threat actor.

Summary

Computer network exploitation has become the primary means of gathering intelligence in today's information age. The proliferation of computer systems in advanced industrial societies represents a boon for would-be cyber spies.

APT is a targeted attack that aims to attack government organizations and business entities; the APT attack is complex and requires sophisticated skills and plenty of resources to carry out. Many researchers consider the APT attack the most significant risk that faces organizations in cyberspace.

It can be devastating if an APT attack targets any organization. The nature of cybercrime allows adversaries to launch their attacks from any place in the world, and their ability to operate anonymously to avoid prosecution imposes real challenges to any organization utilizing IT systems to run its business.

This chapter offered a theoretical foundation of the APT attack; we will postpone talking about APT defense measures to part three of this book. Meanwhile, in the next chapter, we will remain within the scope of complex cyberattacks and talk about another type: ransomware.

Note

1 Arstechnica, "How 'Omnipotent' Hackers Tied to NSA Hid for 14 Years—and were Found at Last", Accessed 2025-04-02. https://arstechnica.com/information-technology/2015/02/how-omnipotent-hackers-tied-to-the-nsa-hid-for-14-years-and-were-found-at-last

6

Ransomware

Introduction

As the digital revolution continues to evolve, human dependence on computing technology to store and process information increases. Nowadays, most
information is created digitally and stored on digital storage devices for easy
access. The increased reliance on digital mediums to store business-related
data has drawn threat actors' attention to this side.

Ransomware is a type of malware designed to prevent access (through
encryption or other means) to a victim's computing device files until a sum of
money – or ransom – is paid to restore access to hijacked data. Ransomware
falls under the category of Digital Extortion, a type of cybercrime, which
includes any cybercrime that aims to force an individual or organization to
pay in exchange for gaining back access to stolen digital assets or taking
some other actions to attackers' demand.

There are different forms of Digital Extortion techniques such as
Ransomware, fake software (fake antivirus programs), negative reviews,
threatening to disclose security vulnerabilities of victim IT systems (or
confidential information) to the press or to the government in addition to
threatening to attack victim organization online presence using Distributed
Denial of Service attack (DDoS) attack if refused to pay money for the
attackers.

No one is immune to ransomware attacks; both individuals and organizations – of all sizes and across all industries – fall victim to this type
of attack, which shows no sign of a slowdown. For instance, ransomware attacks are growing at an explosive rate. According to *Cybersecurity
Ventures*,[1] global ransomware damage costs are expected to reach $265
billion by 2031. According to the same study, the frequency of ransomware attacks will reach one occurrence every 2 seconds by the end of
2031, making ransomware the fastest-growing cyber threat people face
worldwide.

Ransomware operators do not intend to damage the victim's computing
device; instead, they want it to remain operational to enable the victim to use

it to pay the ransom. Ransom payments are commonly done electronically using cryptocurrencies like Bitcoin, making tracing payments extremely difficult. However, some ransomware strains request payments via other anonymous payment methods like gift cards, wire transfers, or premium-rate text messages. The worst side of the ransomware attack model is that the victims are coerced to pay a ransom without any guarantee to restore their locked files.

This chapter will talk about ransomware, its types, delivery methods, infection symptoms, famous ransomware families, and attack phases. We will end by talking about the available options if we fall victims to a success-ful ransomware attack. However, before we begin, let us give a brief back-ground history of ransomware attacks and their evolution.

History and Evolution of Ransomware

The first recorded ransomware attack took place in 1989 when a biologist named "Joseph L. Popp" mailed 20,000 infected floppy disks to the World Health Organization's AIDS conference attendees. These disks were named "Aids Information Introductory Diskettes" and contained a questionnaire. Upon opening it, ransomware called "AIDS Trojan" was triggered to launch after approximately 90 reboots of the victim's computer. The AIDS malware was designed to conceal all directories and encrypt file names located on the **C:\drive**. To unlock victim files, AIDS Trojan demands $189 that should be paid to a Panamanian post-office box. AIDS Trojan used symmetrical encryption to encrypt victim file names and was relatively easy to overcome. However, it lays the stage for what has now become the most devastating threat facing IT systems.

In 1996, two cryptographers, *Adam L. Young* and *Moti M. Yung*,[2] suggested a new type of cyberattack where criminals use asymmetrical encryption to lock victim files – using an encryption malware – and demand payment (ransom) to give the decryption key to restore access. This scenario was hor-rifying and did not take much time to become a reality. The researchers pro-pose the terms *cryptoviral extortion* and *Cryptovirology* to name this type of cyberattack.

Things remained calm till the year 2005 when ransomware utilizing asym-metrical encryption began to appear. The most notable one was Krotten and GPCode; the last one, "GPCode" was the most noticeable as it uses 660-bit RSA encryption, which was considered strong at that time.

In 2006, *Archiveus* ransomware was unleashed, targeting mainly Windows OS users. This type uses strong encryption (RSA). It works by encrypting everything in **My Documents** directory and demanding victims purchase items from three different online pharmacy stores to receive the 30-digit' password to unlock hostage files. *Archiveus* ransomware was mainly spread via SPAM emails and file-sharing websites.

In 2007, the locker ransomware-type (which does not involve encryption) appeared on the scene again. The most famous one was *WinLock*, which works by displaying pornographic images on the victim's screen and demanding them to send a $10 premium-rate SMS to receive the unlock code. *WinLock* spread mainly in Russia and forced most victims to format and reinstall their OS even after paying the ransom.

In 2008, another variant of *GPCode* appeared and was named "*GPcode. AK*"; this variant uses more robust encryption (1024-bit RSA), making it harder to crack compared with the previous version. However, the noticeable event that year was not ransomware. It was the introduction of Bitcoin cryptocurrency, which offers an anonymous payment method. The introduction of Bitcoin has impacted the entire ransomware model and encouraged more threat actors to enter the stage and develop more devastating ransomware because there is now a reliable, secure, and anonymous way to extort money from their victims without the fear of being caught by authorities.

Pushed by widespread usage of Bitcoin currency, 2011 has witnessed an explosive growth in ransomware attacks. According to *KnowBe4*,[3] about 60,000 new ransomware samples were detected in Q3 2011, and this number tripled in 2012 to reach 200,000 new samples.[4]

More advanced ransomware strains began to appear in 2013 through 2015. Examples include *Cryptolocker, Torrentlocker, Cryptowall*, and *Teslacrypt*. 2015 also witnessed the emergence of many Ransomware-as-a-Service (RaaS) strains like *TOX, Fakben, and Radamant.* Many 2015 ransomware strains use strong encryption (AES and RSA-2048 bit), making recovering encrypted files almost impossible; this fact leads to a huge increase in ransom payment to reach more than $325 million by late 2015.

In 2016, the first JavaScript Ransomware-as-a-Service was discovered; this allows one ransomware type to infect multiple operating system types (e.g., MacOS and Linux). During this year, many advanced ransomware strains were discovered, such as *Locky, Petya, CryptXXX* (which also steals your Bitcoins), and *SamSam*. These types use strong encryption, and some offer different payment methods – other than cryptocurrencies – to pay the ransom.

In 2017, the *WanaCry* crypto-ransomware attack was the most noticeable and drew the world's attention to the ransomware problem. Although the *WanaCry* attack has only procured about $90,000 for its operators, however, the volume of infection which reach more than 150 countries has resulted in about $4 billion of wastage expense across the globe.[5]

In 2018, the wave of ransomware continued to hit; however, the most noticeable one was *GandCrab,* which spread using the Ransomware-as-a-Service model and was the first ransomware that adopts the agile software development approach. *GandCrab* infects about 50,000 computers, most of them in Europe, and demands a ransom that ranges between $400 and $700,000 in DASH cryptocurrency.

In 2019, ransomware became more mature and began to use new techniques to infect and spread in addition to improving its propagation capabilities. Some ransomware strains attack victim cloud service and encrypt everything there; other strains delete all backup files – and any file that has any backup extension – in addition to volume shadow copies (which is a technology developed by Microsoft to take restorable snapshots of a partition). Popular ransomware strains that appeared – and resurfaced – at that year were: *CryptoMix, GandCrab, LockerGog*a, *vxCrypter, MegaCortex,* and *Ryuk.*

In 2020, ransomware operators shifted their focus toward employing stealing mechanisms to steal victims' data and then threatened to publish it publicly or handle it to other threat actors if their demands were not met. This type of malware is known as *Leakware* and is considered a ransomware variant. *Maze* ransomware is a type of Leakware that appeared in January 2020. The FBI has issued a warning to US companies about this attack.[6] In addition to using Leakware, ransomware operators have improved their attack methods by targeting Network Attached Storage (NAS) Devices, which host victims' backup data to force them to pay the ransom after losing the ability to recover from a backup.

Ransomware attacks continued to rise during the COVID-19 pandemic; for instance, cybercriminals use the COVID-19 theme to lure their victims with phishing schemes. On the other hand, the huge – and rapid – shift of the workforce to remote work has also increased ransomware attacks significantly, as individual computing devices and home networks do not have the same security defenses as corporate endpoint devices and networks. The fast spread of COVID-19 did not give the security industry enough time to develop secure work-from-home models. This fact was well exploited by cybercriminals to target employees working from home to gain a foothold on the internal corporate network.

No one can foresee the future precisely, especially when talking about the ransomware threat; however, it is beneficial to look at some future ransomware statistics to predict future trends and estimated wastage expenses.

- 85% of ransomware attacks target Windows systems.[7]
- The average ransomware payment in 2021 increased by 82% year over year to $570,000.[8]
- The individual ransom of 1400 clinics, hospitals, and other healthcare organizations varied from $1600 to $14 million per attack.[9]
- The healthcare industry is predicted to spend $125 billion on cybersecurity from 2021 to 2025.[10]

- Global cybercrime will cost the world $10.05 trillion annually by 2025.[11]
- 91% of cyberattacks begin with a spear-phishing email.[12]
- 95% of All Ransomware Payments Were Cashed out via BTC-e Platform. Although this study was conducted in 2017, BTC is still the preferred payment method for ransomware operators until now.[13]
- The US Financial Intelligence Unit (FinCEN) identified approximately $5.2 billion in outgoing BTC transactions potentially tied to ransomware payments in 2021.[14]

These were some statistics about current and future ransomware expected damage. As we note, ransomware operators prefer:

- Bitcoin to receive ransoms
- Mainly targeting Windows-based devices,
- Utilize phishing email as the primary attack vehicle to infect target computing devices with ransomware.

Not all ransomware strains extort money using the same method. In the next section, we will differentiate between the two main types of ransomware.

Ransomware Types

Ransomware can be classified into two types:

A. *Encrypting (Crypto) ransomware*: This type is the most prevalent. It uses advanced encryption algorithms to encode victim files and demand a ransom for the decryption key. Strong ransomware uses military-grade encryption to encrypt infected device data. Some ransomware strains scramble infected device file names, making the victim unable to recognize how many files have been affected by the attack.

B. *Locker ransomware* (also called *Computer Locker*): This type does not encrypt infected device files. Instead, it locks the victims from using infected computing devices by preventing them from accessing the desktop and running or accessing any programs or other network services; it then demands a ransom to unlock the device. Most locker ransomware types are – relatively – easy to remove by an experienced computer user without paying the ransom.

There is a new variant of ransomware that has begun to make noise lately, it is called Leakware (sometimes called Doxware). From a technical perspective, we can consider Leakware a ransomware type, as it uses the same techniques employed by ransomware to attack and spread the infection; however, the difference lies in how threat actors behave after successfully infecting the target system. For instance, Leakware steals victim sensitive files (e.g., personal photos, videos, confidential information, chat conversations, emails, trade secrets, and sensitive financial documents if the target is a corporation) and threatens to publish the leaked information publicly if the victim refuses to pay the ransom. This type of attack is more dangerous than ransomware, as the victim has no way to escape from it. Compared with ransomware, victims' files can be restored by using a backup or by simply formatting the infected device's hard drive. Still, in the case of Leakware, stolen information can contain embarrassing and confidential information that brings reputational – and even financial and legal – damage to individuals and businesses. Lately, we have begun to see hybrid attacks that combine both ransomware/leakware in which victim data is first stolen, then encrypted. If the victim refuses to pay the ransom, attackers threaten them to release the stolen data to the public.

> **Note!** Scareware – which includes fake antivirus and other software that claims their ability to fix operating system errors (e.g., fake cleaning software) – also falls under the Digital Extortion category of cybercrime, just like ransomware. Scareware operators trick unsuspecting users into believing that their devices have malware or other technical problems, such as registry errors, and encourage them to purchase these tools (fake software) to fix the problem. Such applications are not functioning correctly, and many of them contain spyware used to gather sensitive information. Total Antivirus 2020 is an example of a fake antivirus product that employs scare tactics to convince users to purchase it.

Ransomware Infection Symptoms

Different indications appear when falling victim to ransomware; the following list the most prominent ones:

A. *Ransom note*: Splash screen pop-up in front of the computer screen announcing you are a victim of a ransomware attack and displaying detailed instructions on how to pay the ransom (see Figure 6.1).

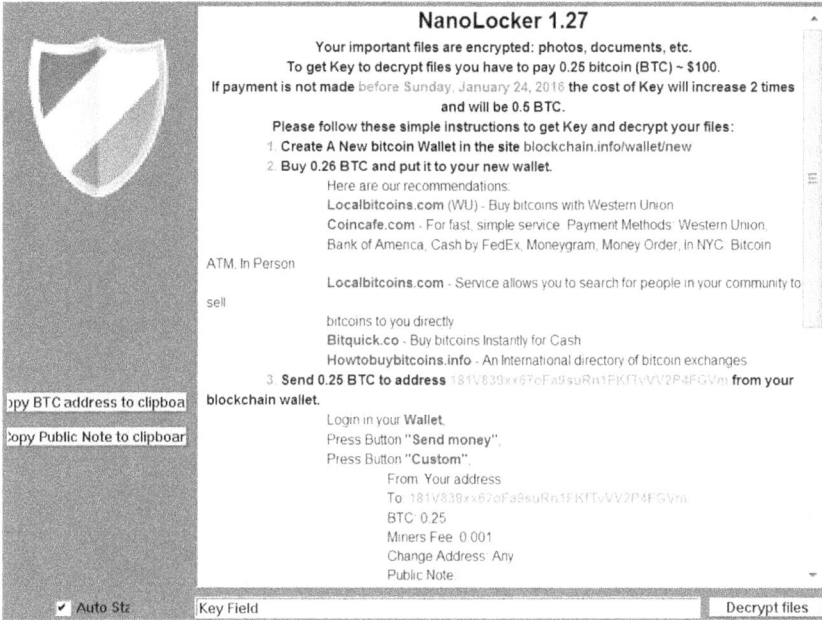

FIGURE 6.1
Sample ransomware screen notice.

B. *Files are not open*: You will notice your files –and even some programs, especially web browsers – do not open when trying to open them. If you are using Windows 10, the operating system will launch an error message *"Windows cannot open this type of file"* and asks you to choose a program to open the intended file type (see Figure 6.2).

C. *Scrambled filenames*: Some ransomware varmints scramble original filenames and replace them with a random file name and extension. *CryptXXX* is an example of ransomware that scrambles infected device filenames.

D. *The existence of instruction files*: After infecting the victim device with ransomware, the ransomware will leave files (usually text and HTML files) containing information on how to pay the ransom. Such

FIGURE 6.2
Windows 10 error message when trying to open an unknown file extension.

Man in the middle attack

Original Connection

User

Web Application

Hacker

Man in the Middle

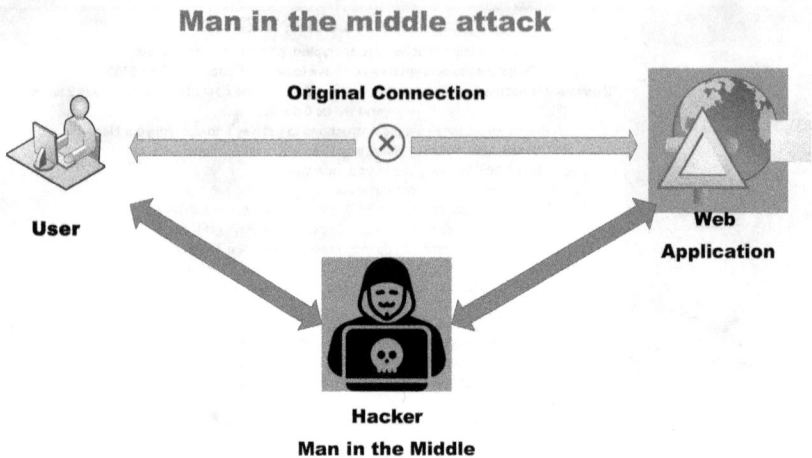

FIGURE 6.3
Sample ransom note left by the THANOS ransomware.

files can be found on the infected device desktop and sometimes left on every folder containing encrypted data. The instruction file names are commonly written in all caps letters to draw the victim's attention, such as "README_FOR_DECRYPT", "DECRYPT YOUR FILES", and "HOW_TO_DECRYPT_FILES". See Figure 6.3 for a sample ransom note of the THANOS ransomware.

Ransomware Delivery Methods

Ransomware is a type of malware that shares the same delivery methods as other malware types. In this section, we will mention the primary delivery methods employed by ransomware to spread and infect computer systems.

1. Email attachments & malicious URLs within the body of the email and social media messages
2. Exploit kits
3. Malvertising
4. Infected USB devices
5. Free internet software downloads.
6. Cracked programs infected with ransomware and downloaded from pirated software websites.
7. Ransomware as a service (RaaS)

Exploiting the Remote Desktop Protocol (RDP) and Managed Service Providers. We covered each delivery method in detail in Chapter 4 when discussing cybercrime attack types and attack vectors. However, we did not discuss utilizing ransomware-as-a-service and exploiting RDP/MSP services attack vectors, so we will discuss them in detail in this section.

Ransomware-as-a-service

Ransomware has proved to be a profitable criminal business model for cyber-criminals. This encouraged them to expand their operations by offering attack platforms for spreading ransomware by other malicious actors who do not possess the required technical skills to launch ransomware attacks on their own.

Similar to the software-as-a-service (SaaS) concept offered by many giant tech vendors, where software licenses are provided on a subscription basis and hosted centrally on the vendor's cloud servers. Ransomware as a service (RaaS) (see Figure 6.4) plays the role of a broker used by RaaS operators. RaaS operators sell ready-to-launch ransomware attacks for other junior attackers (affiliates) in exchange for a percentage of the acquired ransom (commonly

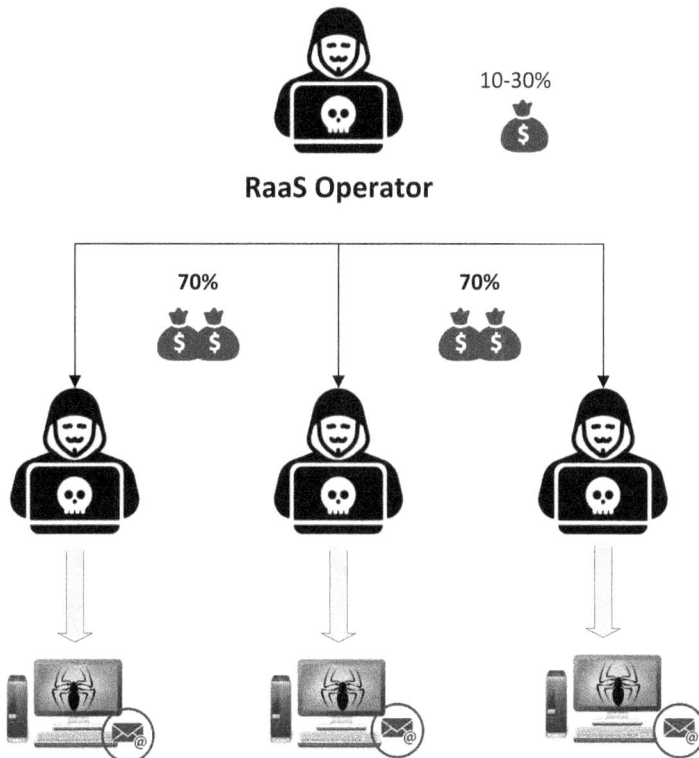

FIGURE 6.4
How RaaS business model works.

between 1% and 30% of the acquired ransom). RaaS operators are responsible for upgrading the ransomware and ensuring it bypasses security defenses to maintain its value in the cybercriminal world.

Cybercriminals deploy RaaS on the darknet, especially on the TOR anonymous network, to make tracking their activities – and payments – extremely difficult. There are many RaaS platforms on the darknet, and they are easy to use by invoice attackers. Each platform comes supplied with a dashboard to manage clients (victims) and monitor received ransom payments.

The most notable ransomware strains distributed using the RaaS model are *Satan, Cerber, Philadelphia,* and *Sage.*

Exploiting Remote Desktop Protocol

Remote Desktop Protocol (RDP) is a proprietary protocol developed by Microsoft Corporation, available in most Windows versions, and uses TCP port 3389 and UDP port 3389 by default (also available in other operating system types such as Apple and some versions of Linux). RDP is used for remote connection, it allows a user to connect to another computer or server – via the internet or other network connection – and control it as if it was attending it in person (both devices should support RDP for the connection to work).

RDP is commonly used in the corporate environment to simplify accessing corporate resources (files and applications) from remote locations. However, due to COVID-19, most organizations worldwide have adopted the work-from-home model to remain operational during the ongoing pandemic. This quick – and emergency – transition for telework did not allow most organizations to configure their RDP connection properly for security. This situation offers a unique opportunity for ransomware gangs to exploit weaknesses of unsecured RDP connections to infect their targets with ransomware.

Attackers use the brute-force password guessing attack technique to crack RDP passwords. Social engineering tactics are also utilized to acquire RDP account credentials from unaware users. Exposed RDP servers can be found using specialized search engines that look for internet-connected devices like SHODAN. For example, we use SHODAN to search for exposed RDP; the returned result displays different technical information about each device, including operating system type and web technologies. SHODAN also displays the vulnerabilities that could impact the subject device based on the discovered software and version.

According to a report published by Kaspersky,[15] the number of brute-force attacks against RDP internet-facing servers has increased from 200,000 per day in early March 2020 to over 1,200,000 during mid-April 2020. The RDP brute-force attacks have accelerated further in the last four months of 2021 to reach 206 billion, according to ESET Threat Report T2 2021.[16]

RDP is considered secure when connecting two devices within a private network; however, security concerns arise when using it to communicate through the internet.

Note! Attacks against RDP have accelerated during the COVID-19 pandemic due to increased work-from-home situations. **Stolen RDP credentials** are sold on the darknet for a few dollars for each breached account.[17]

Exploiting Managed Service Providers

A managed service provider (MSP) is an IT vendor that provides different remote IT services (on a subscription basis) to other organizations. MSP services include IT support, network monitoring, customer invoicing and online bill payment, software applications, IT infrastructure management, and other IT consultation services. MSP customers are commonly small to medium-sized enterprises that want to outsource some of their IT work to reduce costs and increase work efficiency.

An MSP provider uses a specialized remote management tool (similar to RDP) to manage the IT assets of the client remotely. The fact that MSP providers can have broad access to their customers' IT environment makes them an attractive target for ransomware gangs to exploit their service and infect MSP clients.

The *GandCrab* and *BitPaymer* ransomware strains are known to spread using this method. In the future, the MSP work model is expected to gain more popularity making it occupy an important position among other ransomware attack vectors.

Ransomware Attack Phases

Ransomware passes five phases before announcing its presence on the victim's device. Knowing what happened in each phase helps you to recognize the signs and work promptly to defend your business against an attack or at least to lower its effects.

1. *Deployment & Infection*: In this phase, the malware enters the system via a spam/phishing email, an exploit kit, a malicious USB drive, or any of the attack vectors already mentioned.
2. *Execution*: Now that the malware has successfully entered the victim's system, the actual payload of the ransomware is downloaded from the internet and executed to infect the system. The persistence mechanism of the ransomware is established in this phase.
3. *Backup Destruction*: In this phase, the ransomware searches for available backup within the infected system and all other attached

backup drives (e.g., external hard drive, USB memory) in addition to other accessible network shares – if the backup is stored across the network – and destroys them to prevent any recovery. Please note that not all ransomware strains destroy backups on the infected system; however, modern strains do this generally.

4. *Encryption (see* Figure 6.5*)*: After destroying the infected system backup files, the ransomware will communicate with its command and control server (C&C) and create an encryption key pair to be used on that device. Some ransomware strains can do this locally without connecting to the attacker's remote server. Ransomware encrypts all files within the infected computer and all attached drives and network shares to which the computer has access.

5. *Notification*: After removing backup files and encrypting the infected system files, a ransom note appears on the victim screen with detailed instructions on how to pay it. Generally, three to seven days are given for the victim to pay. After that, the ransom increases. If no payment is made in the specified period, complete destruction of the

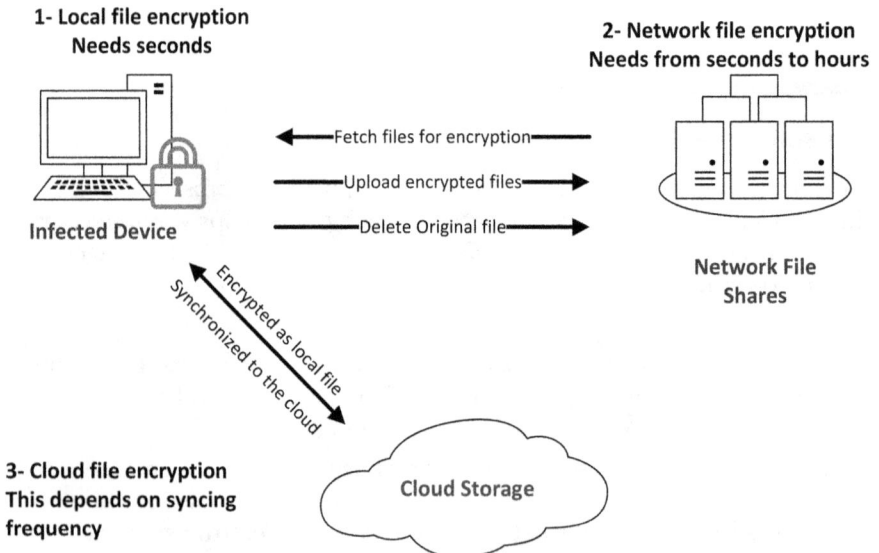

Ransomware Encryption Process
Encryption always happens on the infected device

1- Local file encryption
Needs seconds

2- Network file encryption
Needs from seconds to hours

←—————Fetch files for encryption————

—————Upload encrypted files—————→

—————Delete Original file—————→

Infected Device

Network File Shares

Encrypted as local file
Synchronized to the cloud

3- Cloud file encryption
This depends on syncing frequency

Cloud Storage

© www.DarknessGate.com 2021

FIGURE 6.5
Ransomware encryption process.

encrypted data will be performed. Advanced ransomware strains have a self-destruct capability to prevent any forensic analysis of their artifacts that can help create better defenses against them in the future.

Popular Ransomware Families

Ransomware can be classified according to different criteria. The general method is to classify ransomware according to its code signature, which identifies the malicious code and its behaviors. In this section, we will discuss the most prominent ransomware variants according to their release date. In the last section of this chapter, we will mention the available decryption tool for each ransomware variant.

Conti Ransomware

The most recent trends that ransomware gangs began to follow were:

- Creating cross-platform ransomware
- Ransomware gangs start to organize and evolve their work as legitimate enterprises

Creating cross-platform ransomware is a recent trend that appears more frequently in 2021-2022. Threat actors begin to develop cross-platform ransomware to increase their profits when attacking complex IT environments that include different operating systems and architectures. Cross-platform ransomware is written using a cross-platform programming language, such as Rust or Golang, so it can run on different operating systems such as Windows, Linux, and Mac.

By using ransomware executable on different architectures, the threat actors can:

1. Encrypt a large number of system files belonging to different operating systems to make recovery very difficult and time-consuming, and hence, force victims to pay the ransom quickly
2. The same ransomware can be easily ported to another platform if threat actors change their minds and aim to target another entity that is using a different type of OS
3. Analyzing the source code of cross-platform binaries is relatively more complex than those written in the C programming language. This increases the overhead when trying to disassemble the ransomware source code

The most notable ransomware group that utilizes cross-platform ransomware is named Conti. Security researchers and people who work in the intelligence areas think the Russian government sponsors the Conti ransomware group. In February 2022, leaked chats[18] between Conti group members show their support for the Russian government during its invasion of Ukraine. The leaked chat messages revealed important facts about the workings of this criminal group. For instance, their management hierarchy resembles those that existed in regular tech organizations. The group's annual payroll reaches $6 million paid for a group of between 65 and 100 hackers. The exciting thing about this group is that it has a bounce system for its hackers, a performance review, training opportunities, and even an "employee of the month" reward! Conti acquired about $180 million of ransom from data extortion activities in 2021.[19]

The primary attack vector used by the Conti group is spear-phishing emails and tech support fraud – where hackers impersonate known companies and provide services to fix computer problems over the phone or email. Conti gathers intelligence information about its targets by buying leaked information from other criminal groups and using it to customize their attacks.

Conti buys security solutions software products from legitimate commercial software providers to test their malware before launching their attack; they also purchase ready-to-launch exploits and attack tools from other cybercriminals.

The FBI[20] warned in its "Internet Crime Report 2021" that the Conti group was among the three top variants that targeted critical infrastructure in the United States in 2020. Conti is considered among the most successful ransomware groups because it successfully achieved $2.7 billion in cryptocurrency in only two years.

Maze

This is a modern encryption ransomware that targets Windows-based devices. It first appeared in the first half of 2019 and aims to target big organizations. Some of its victims include: *LG, Southwire,* and the *City of Pensacola. Maze* ransom note consists of the title *"0010 System Failure 0010"*, and many security researchers reference it using the name *"ChaCha ransomware"*.

Like most Ransomware, *Maze* spreads via malicious attachments (e.g., malicious MS Office macros), SPAM emails, and exploit kits (especially *Fallout EK* and *Spelevo EK*). The threat actors behind *Maze* ransomware utilize different attack vectors such as spear-phishing campaigns, exploiting weak RDP passwords, and vulnerable VPN services. They also use attacks and reconnaissance tools such as Mimikatz, ProcDump, Cobalt Strike, Advanced IP Scanner, Bloodhound, and PowerSploit to compromise the target network.

Maze threatened its victims to reveal their sensitive data if they refused to pay the ransom. This makes it play the role of Leakware. *Maze* threat actors maintain a website on the TOR network where they publish their victims' leaked information if they refuse to cooperate.

In June 2020, *Maze* operators teamed up with two other ransomware groups (*LockBit* and *RagnarLocker*) to form a ransomware cartel. They exchange their attack techniques and publish the exfiltrated data on the Maze website.

Ryuk

Ryuk is a crypto-ransomware attributed to the *APT Lazarus* hacking group, which is thought to be sponsored by the North Korean intelligence services. However, some researchers found clues that *Ryuk* could be linked to Russia. It first appeared in 2018. *Ryuk* mainly targets high-profile enterprises (organizations that work in Technology, Healthcare, Energy, Financial Services, and the Government sector) for a high ransom return (in some cases, *Ryuk* ransom demand exceeds USD 300,000[21]). After analyzing its source code, code experts found many similarities between *Ryuk* and the *Hermes* ransomware. This makes them believe that *Ryuk* is derived from it. However, unlike *Hermes,* which was used by many threat actors and its source code was published for sale on many darknet websites, *Ryuk* was only used by the *APT Lazarus* group to use it in their targeted attacks.

What makes *Ryuk* attack dangerous is its dependence on pentesting and open-source toolkits like *Mimikatz, PowerShell, PowerSploit, LaZagne, AdFind*, and *PsExec* in addition to being manually operated and leveraging a multi-stage attack. For instance, *Ryuk* preceded its attack by infecting with two malware *Emotet* and *TrickBot,* so if an organization detects *Ryuk* infection, then there is a high probability that other malware existed in its system.

Ryuk spread mainly via phishing emails. After infecting the victim device, *Ryuk* leaves a read me file named *RyukReadMe.txt* on each folder containing encrypted files. This file contains information to pay the ransom using Bitcoin. The interesting thing with the Ryuk ransom note is that it lists two emails (one on ProtonMail.com and the second on Tutanota.com) to contact the attacker to know the amount of money you should pay to get the decryption key.

Ryuk attacks have increased significantly during 2020. For instance, through Q3 2020, 67.3 million *Ryuk* attacks were detected – 33.7% of all ransomware attacks this year. In 2021, Ryuk ransomware underwent new development. A new variant with computer worm capabilities has appeared, meaning it can travel across computer networks and infect a large number of systems without human intervention.

WannaCry

WannaCry is an encryption ransomware with worm capabilities. It can travel across computer networks rapidly to infect a large number of vulnerable computers in a short time. *WannaCry* gained popularity on May 12, 2017 when a huge attack targeted different countries worldwide. It targets Windows-based devices and demands $300 as a ransom to decrypt the hijacked data. If no payment is made within three days, the ransom is doubled; after one week, the hostage data will be destroyed if the victim fails to pay the ransom.

WannaCry takes advantage of a set of stolen government hacking tools and exploits. For instance, on April 8, 2017, the hacking group Shadow Brokers (TSB) exposed various hacking tools that exploit vulnerabilities in Windows OS and other security software products like firewalls, IDS, and antivirus products. *WannaCry* uses a ready exploit named *EternalBlue* developed by the NSA. This exploit takes advantage of a vulnerability that existed in *Server Message Block (SMB)* protocol, which is used on many Windows versions such as Microsoft Windows Vista SP2, Windows 7 SP1, Windows Server 2008 SP2 and R2 SP1, Windows 8.1, Windows Server 2012 Gold and R2, Windows RT 8.1, and Windows 10 Gold 1511, and 1607, and Windows Server 2016, to spread and infect other unpatched computers on the network.

WannaCry is known to be operated by *Lazarus Group,* which is thought to be connected to the North Korean government. *WannaCry* caused huge global losses that reached 4 billion USD, with more than 230,000 computers infected in 150 countries. Many famous companies get infected with this ransomware, such as *FedEx, Telefonica, Nissan, and Renault.*

Note! To close the SMB vulnerability exploited by *WannaCry*, Microsoft released a hotfix update that can be downloaded from https://technet. microsoft.com/en-us/library/security/MS17-010.

Cerber

Cerber is the most prominent early adopter of the ransomware-as-a-service (RaaS) model. The ransomware developers take about 40% of the ransom as a commission from other affiliated cybercriminals – who do not possess any coding experience – for each successful attack. It first appeared in July 2016. *Cerber* targeted Windows-based devices and used different attack vectors to infect its victims such as: malicious attachments –especially MS Word files with malicious macros, exploit kits, and sometimes disguised in free internet software.

Before infecting the target system, *Cerber* checks its geographical location. If the victim device is from any of the following countries, then no encryption is made: *Armenia, Azerbaijan, Belarus, Georgia, Kyrgyzstan, Kazakhstan, Moldova, Russia, Turkmenistan, Tajikistan, Ukraine, Uzbekistan.* If the victim resides in any other country, *Cerber* begins its encryption using the AES-256 and RSA algorithms to encrypt victim files and changes the infected files' extensions to .cerber. (Modern versions of *Cerber* change the infected file extension to .cerber2, .cerber3, or four random characters). After successfully encrypting the system, the ransomware will drop three files – with different extensions – on the victim's desktop carrying the name "# DECRYPT MY FILES #" containing information about the ransom payment amount and the payment method.

There are many variants of *Cerber* ransomware; the most notable one appeared in May 2016; it comes bundled with a DDoS attack software making victim devices join a network of DDoS bots in addition to infecting them with ransomware. Another variant appeared in 2017 and added techniques to evade detection by security solutions that leverage machine learning technology for detection.

Locky

Another crypto-ransomware first appeared in 2016. *Locky* targets Windows-based devices and uses different infection vectors to spread; the most popular one is using a malicious Microsoft Word document that runs infectious macros. When opened by the user, the malicious file will appear in an unreadable format; a message will pop up asking the victim to enable macros using the following phrase: "*Enable macro if data encoding is incorrect*". If the user enables the macros, then the malicious script will get executed, download the Locky executable file, and store it in the "Temp" folder of the current user account. Once installed, it will begin encrypting data on all local drives, connected USB drives, and remote drives, and deleting Windows *Volume Shadow Copy* files to prevent any file recovery. *Locky* usually demands a ransom of about $400 in Bitcoin cryptocurrency.

Locky targets small and medium businesses and focuses on encrypting specific file types, especially source code files, database files, and files created by designers, developers, and engineers. Security researchers think *Locky* originated from Russia, as it attacks devices located in all countries except Russia or the one with the Russian language installed.

Petya

Another strain belongs to the encrypting ransomware family; two versions of Petya ransomware exist. The first one appeared in 2016; it spread via SPAM emails and behaved like other crypto-ransomware, while the second variant appeared in 2017 and was called *NotPetya*. Although this variant spread and infected similarly to crypto-ransomware, it was considered to belong to the cyber weapon category.

Petya infects Windows-based devices by attacking the Master Boot Record, then executes the payload responsible for encrypting all data on the infected device's hard drive (it specifically encrypts Master File Table (MFT) & Master Boot Record (MBR)) and prevents Windows from loading asking for a

$300 ransom in Bitcoin to remove the restriction. The early version of *Petya* required administrative privilege to work. However, other variants (version two and three) appeared later in the same year (2016) and worked by install- ing another malware called *Mischa* if *Petya* lacks the necessary privilege to access the MBR of the target device's hard drive. Both *Petya* and *Mischa* can work offline without communicating with their C&C servers to receive the encryption key.

In June 2017, another variant of *Petya* occupied headlines (named *NotPetya*), *NotPetya* initiated its presence through a massive attack that hit different Ukrainian targets and later spread worldwide, causing 10 USD billion in losses.

Note! What is Master File Table?

The MFT is a database file that contains information about each file and directory in the Windows NTFS file system. There is at least one record for each file and directory on the NTFS file system partition that contains important information/attributes (time, date stamps, size, and permissions) that tell the operating system how to deal with the file or directory. More information about the NTFS file system structure can be found at: http://ntfs.com/ntfs-mft.htm

---

NotPetya belongs to the cyber weapon category and not cybercrime because of its destructive behavior. For instance, *NotPetya* uses the same ransomware tactics in encrypting infected device files and demanding a ransom; however, this variant generates the encryption key randomly and then destroys them, making decrypting the infected device files back impossible.

The *NotPetya* ransomware was spread via the *EternalBlue* exploit, initially developed by the *US National Security Agency* (NSA) and leaked by the *Shadow Brokers* group. The same exploit was used previously to infect with the *WannaCry* Ransomware.

Several governments around the world accused the *Russian* government of being responsible for *NotPetya* attack because the main target of its attack was the Ukrainian network.

SamSam

SamSam is encrypting ransomware used in targeted attacks; it first appeared in late 2015, however, it's the year 2018 when *SamSam* began to draw attention in media after hitting high-profile targets and government organizations like the *City Of Atlanta*[22] (took place on March 22, 2018 and cost it $2.6 to recover its systems after refusing to pay the $53,000 ransom!),

City of Farmington,[23] *Hancock Health,*[24] and *Allscripts,* a billion-dollar electronic health record (EHR) company.[25] *SamSam* operators select targets likely to pay the ransom to recover their critical data due to their sensitive work, such as hospitals, educational institutions, and public sector organizations.

SamSam operators leverage vast arrays of security tools to discover and attack their targets. Some of these tools include Windows *Sysinternals* toolkits and *Mimikatz* for password cracking. *SamSam* operators work by exploiting vulnerabilities in target networks such as outdated or vulnerable servers, weak passwords used on RDP or VPN accounts (they crack the password using different brute-force techniques), as well as vulnerable FTP platforms and Microsoft's IIS. After that, work to deploy different security tools to discover and map the target network before executing ransomware and beginning data encryption.

Unlike other ransomware strains that need human intervention – such as opening an attachment or visiting a compromised website housing an exploit kit – to execute. *SamSam* does not depend on user action, as attackers can launch the ransomware remotely once found a vulnerable service/application.

In this section, we've mentioned the most prevalent ransomware strains that have appeared since 2015 and talk briefly about each strain and the methods it uses to infect. However, keep in mind, that there are many other ransomware families, and each one has many variants. For example, *TeslaCrypt* was popular in 2015, *CryptoWall* in 2014, and *CryptoLocker* in 2013.

Handling Ransomware Incidents

No one is immune to a ransomware attack. Even the most guarded organization still falls victim to this type of attack. Whether you train your employees to handle phishing emails, avoid visiting malicious sites, and clicking suspicious links, even installing the best security solutions may not be enough to stop this type of attack. It only takes one misstep for the attackers to infiltrate the most secure network and hold your data hostage.

Having a backup is the most effective recovery strategy against ransomware attacks. However, in many scenarios, this option may not be available, leaving the victim organization or individual against three options to choose from:

1. Do nothing and accept the loss of data. Format the infected device and install a new operating system and applications
2. Find a decryption tool to decrypt infected data
3. Pay the ransom

After a successful ransomware attack, the encrypted data is still in possession of the victim, but the cost of rereleasing it can add up quickly. Understanding the value of data to organization work can help it decide which option is the most convenient. For example, when infected data is used in daily work operations, losing access to it can significantly impact work continuity. Examples of such data include patient records, legal documents, source code files, drawing files (e.g., AutoCAD files), and training materials, to name a few. The estimated damage cost is not only related to the loss of revenue because of work disruption, but reputation and brand damage can also be enormous, and sometimes its cost cannot be measured.

Once an organization understands how to measure the significance of its data, it can now select the most appropriate resolution to a ransomware attack. In the following lines, we will cover each option in some detail.

Do Nothing!

This is the easiest – and cheapest – option. In such a scenario, the victim will wipe – format – infected device storage media, including installed applications and the operating system, and install a fresh copy. This option may not always be feasible as encrypted data may contain critical data for organization work. However, this is the only solution if the victim is unwilling to pay the ransom and no decryption tool is available to decrypt the infected data.

Many organizations select this option when infected data has minimal value for business work, or data can be replaced easily without downtime or paying any extra fees.

Remember to format all partitions of the infected hard drive when following this option. For instance, some ransomware strains may plant additional malware on all partitions of the infected device's hard drive, which can be executed at any time to repeat the infection.

> Note! Before wiping hard drives containing infected data, it is advisable to keep a copy of encrypted data in case a decryption tool becomes available in the future.

Using a Decryption Tool

This option depends on whether there is a decryption utility for the specific ransomware. In all cases, when falling victim to a ransomware attack, it is advisable to search for a decryption tool for that type of ransomware. Even if you have a backup of the infected data, the availability of a decryption tool can speed up the recovery process.

As ransomware risk intensifies, security firms are investing in creating decryption tools for various ransomware families. Soon after a ransomware strain hit the landscape, security vendors and independent researchers worldwide work to create a decryption tool to circumvent it.

Remember that a decryption utility is designed to decrypt one type of ransomware strain and not all types. Besides, such tools are not 100% foolproof. However, many of them gave excellent results in restoring locked data.

Before we list the various websites where you can find ransomware decryption tools, you should first know how to identify the type of ransomware strain that has infected your computer.

Identify Ransomware-Type

There are various methods to discover the ransomware strain that infects your system. The simplest method is reading the instruction file left by the ransomware or the ransom note. There is a high probability that a ransomware name is mentioned in such places.

The following are some online services for identifying ransomware strains:

1. *ID. Ransomware* (https://id-ransomware.malwarehunterteam.com): This is a free online service for identifying the ransomware type that infects your system. Currently, it detects more than 1068 different ransomware. There are various methods to use this service; for instance, you can upload the ransom note, upload an encrypted sample file (one of your encrypted files), supply the email address/s, or the links mentioned in the ransomware instructions file to determine the type of ransomware strain. If there is a decryptor for the identified ransomware strain, the ID ransomware website will give a link to download it.

2. *No More Ransom* (https://www.nomoreransom.org/crypto-sheriff. php): Another free online service for detecting ransomware-type. To use this service, you can either upload an encrypted ransomware file, or supply the information found in the ransomware note such as email, website URL, onion, and/or Bitcoin address.

Now that we can identify the type of ransomware that hit our system, let us move to the next step to see where we can find a tool to remove the ransomware restriction.

Ransomware Removal Utilities

As we already mentioned, ransomware is branched into two types: locker and encryption ransomware. Encrypting ransomware – which is the most common – uses encryption algorithms to encrypt infected device files. The locker ransomware does not encrypt victim files. Instead, it locks the infected device's operating system and prevents users from accessing computer files or applications.

Remove Screen-Locking Ransomware

Locker ransomware can be removed manually by using any of the following two methods.

Boot into Safe Mode You can try removing screen-locking ransomware by rebooting your computer into Safe Mode and removing the malware using a reliable antivirus solution. When booting into Safe Mode, Windows will only load the necessary drivers and applications to start appropriately and not start the locker ransomware. A user can then download and execute an antivirus or use the existing one to remove the ransomware. Microsoft has a guide about booting into Safe Mode for Windows 10 devices.[26]

Use System Restore Feature Another manual method is utilizing the Windows Restore Point feature to recover your computer to the last well-known state. *How to Geek* offers a detailed guide on restoring Windows 7, 8, and 10.[27]

Note! *Trendmicro* has a guide on removing lock-screen-type ransomware on Android devices (https://helpcenter.trendmicro.com/en-us/article/TMKA-01028).

Ransomware Decryption Tools

Using some ransomware decryption tools can be straightforward. However, some decryptors require technical knowledge to execute them. This section will mention sites where you can find ransomware decryptors. Keep in mind that the list of decryptors is changing every day, so if you cannot find a decryptor for a specific ransomware, try to conduct a Google search to see if one has been developed lately.

1. *Emsisoft* (https://www.emsisoft.com/ransomware-decryption-tools/free-download): This company offers scores of free ransomware decryption tools.
2. *Nomoreransom* (https://www.nomoreransom.org/en/decryption-tools.html): List various ransomware decryption tools along with each removal guide created by the tool developer. This website should be your first option when searching for ransomware decryptors. *No More Ransom* is a project developed by the National High Tech Crime Unit of the Netherlands' police, Europol's European Cybercrime Centre, Kaspersky, and McAfee to find a ransomware solution for victims to recover their locked files without paying a ransom.
3. *Kaspersky* (https://noransom.kaspersky.com): The *Kaspersky No Ransom* project lists many free ransomware removal tools.

4. *Trend Micro* (https://success.trendmicro.com/solution/1114221-downloading-and-using-the-trend-micro-ransomware-file-decryptor): *Trend Micro Ransomware File Decryptor* tool decrypts files infected with many ransomware families.

5. *McAfee* (https://www.mcafee.com/enterprise/en-us/downloads/free-tools/ransomware-decryption.html): McAfee Ransomware Recover (Mr2) tool can unlock infected files with many ransomware types.

6. *Bitdefender* – GandCrab Ransomware decryption tool (https://labs.bitdefender.com/2018/10/gandcrab-ransomware-decryption-tool-available-for-free) and Bart Ransomware Decryption Tool (https://labs.bitdefender.com/2017/04/bart-ransomware-decryption-tool-released-works-for-all-known-samples).

7. *Avg* (https://www.avg.com/en-us/ransomware-decryption-tools): AVG has many free ransomware decryption tools.

8. *Heimdalsecurity* (https://heimdalsecurity.com/blog/ransomware-decryption-tools): This website lists links to scores of free ransomware decryptors offered by many security companies and other independent researchers.

9. *Knowbe4* (https://blog.knowbe4.com/are-there-free-ransomware-decryptors): List 100+ free ransomware decryption tools.

Paying the Ransom

Paying a ransom is the worst option, which we do not recommend personally! There is no guarantee to restore access to your encrypted data after paying the ransom. On the other hand, it is not good to negotiate with cybercriminals. Such people lack the ethics to honor their word. Besides, paying money to ransomware operators will encourage them to develop more sophisticated malware to target more people.

Despite everything, losing access to data in today's information age can have catastrophic consequences on victim organizations. Consider a hospital exposed to a ransomware attack that encrypts all patients' records; how can things go in such a scenario? Is there a threat to patients' lives if their medical information becomes inaccessible? For business organizations, ceasing operations can cost a lot of money and reputation loss; it may even lead to a complete shutdown of the victim organization with all the social consequences resulting from making hundreds or even thousands of people unemployed.

Therefore, paying a ransom can be the only solution to a big problem. We are not going to discuss how someone can pay a ransom. However, as we already mentioned, ransomware operators prefer cryptocurrencies – especially Bitcoin – to receive their ransom funds. The ransom payment is commonly conducted via the TOR anonymous network.

To pay a Bitcoin ransom, you need to have a Bitcoin account funded with Bitcoin currency. In Bitcoin naming, the account number is donated as an *address* (A Bitcoin address comprises 26-35 alphanumeric characters and looks similar to this: 19r9k2Rq2PZhGRc6dsi3j1KQ7fJEZwbcVe). A single user can have many addresses associated with his account; however, all these addresses are stored in one place called a Wallet in Bitcoin naming.

The Bitcoin wallet is a software program (sometimes comes in the form of a hardware device) installed on your computer or an application on your smartphone and is used to generate Bitcoin addresses. A wallet is a virtual location where your Bitcoin cryptocurrency fund is stored. You can use it to send/receive Bitcoin in addition to tracking previous transactions.

There are many online services for creating a Bitcoin wallet. The following list is the most popular:

1. https://wallet.bitcoin.com (official Bitcoin wallet).
2. https://www.blockchain.com/wallet
3. https://www.exodus.io
4. https://electrum.org
5. https://wallet.mycelium.com – for mobile devices
6. https://trezor.io – a hardware wallet

Suppose you plan to store some money in your Bitcoin wallet. In that case, it is advisable to use a software-based wallet instead of an online service, as online wallet services have suffered many security breaches.

> Note! Although it is unlikely for modern ransomware strains, some ransomware-type request payments via gift cards such as Amazon and iTunes gift cards.

We will not describe how to conduct Bitcoin transactions as this is outside book scope; however, if you want to know this process in detail, refer to the authors book *Ransomware Revealed: A Beginner's Guide to Protecting and Recovering from Ransomware Attacks* published by Apress 2019 – Chapter 7.

To conclude this section, each ransomware incident is unique; we cannot propose a single solution to all ransomware attacks. If you fall victim to a ransomware attack, do not panic and seek help from a professional fellow, and make sure to report the incident to authorities. You should also conduct an online search after discovering the type of ransomware infecting your system, as there is a chance to find a decryptor to recover the hostage data.

Summary

Ransomware has proven to be an essential form of cyberattack that infects all organizations and individual users. This chapter introduces the ransomware threat, discusses its types, infection symptoms, distribution methods, and attack phases, and gives a brief history of its evolution.

We also talked about famous ransomware families and concluded this chapter by discussing ransomware incidents and our options if we get infected with ransomware. It is a tough decision for any organization fall victim to a ransomware attack on how to behave. For organizations that do not have in-house security professionals, it is advisable to seek advice from a consulting security firm to decide the best action specific to the organization.

It is always advisable not to pay the ransom, even though there will be a disturbance to your business. There is no guarantee that ransomware gangs will respect their word and give you back the decryption key after paying the ransom.

The next chapter will discuss the most important cybercriminals' methods to gain unauthorized access to computer systems. Social engineering is the psychological manipulation of people to convince them to reveal sensitive information, mostly account credentials, to gain entry to the target IT environment and computing devices. Most sophisticated cyberattacks like APT and ransomware attacks take advantage of SE to gain a foothold in the target network. SE will be covered in detail in the coming chapter.

Notes

1 Cybersecurityventures, "Global Ransomware Damage Costs Predicted to Exceed $265 Billion By 2031", Accessed 2025-04-02. https://cybersecurityventures.com/global-ransomware-damage-costs-predicted-to-reach-250-billion-usd-by-2031
2 Adam Young and Moti Yung, "Cryptovirilogy: Extortion-Based Security Threat and Countermeasures", Proceedings of the 1996 IEEE Symposium on Security and Privacy. Accessed 2025-04-02. https://www.ieee-security.org/TC/SP2020/tot-papers/young-1996.pdf
3 Knowbe4, "Ransomware", Accessed 2025-04-02. https://www.knowbe4.com/ransomware
4 McAfee, "McAfee Labs Threats Reports", Accessed 2025-04-02. https://www.mcafee.com/enterprise/en-us/threat-center/mcafee-labs/reports.html
5 Kaspersky, "What is WannaCry ransomware?", Accessed 2025-04-02. https://www.kaspersky.com/resource-center/threats/ransomware-wannacry
6 Cyberscoop, "FBI Warns U.S. Companies about Maze Ransomware, Appeals for Victim Data", Accessed 2025-04-02. https://www.cyberscoop.com/fbi-maze-ransomware

7 Safetydetectives, "Ransomware Facts, Trends & Statistics for 2022", Accessed 2025-04-02. https://www.safetydetectives.com/blog/ransomware-statistics

8 Purplesec, "2024 Cybersecurity Statistics", Accessed 2025-04-02. https://purple-sec.us/resources/cybersecurity-statistics

9 Purplesec, "2024 Cybersecurity Statistics", Accessed 2025-04-02. https://purple-sec.us/resources/cybersecurity-statistics

10 Cybersecurityventures, "Ransomware Runs Rampant On Hospitals", Accessed 2025-04-02. https://cybersecurityventures.com/ransomware-runs-rampant-on-hospitals

11 Cybersecurityventures, "Cybercrime to Cost the World $10.5 Trillion Annually by 2025", Accessed 2025-04-02. https://cybersecurityventures.com/hackerpocalypse-cybercrime-report-2016

12 Knowbe4, "91% of Cyberattacks Begin with Spear Phishing Email", Accessed 2025-04-02. https://blog.knowbe4.com/bid/252429/91-of-cyberattacks-begin-with-spear-phishing-email

13 Bleepingcomputer, "95% of All Ransomware Payments were Cashed Out Via BTC-e Platform", Accessed 2025-04-02. https://www.bleepingcomputer.com/news/security/95-percent-of-all-ransomware-payments-were-cashed-out-via-btc-e-platform

14 Fincen, "Ransomware Trends in Bank Secrecy Act Data Between January 2021 and June 2021", Accessed 2025-04-02. https://www.fincen.gov/sites/default/files/2021-10/Financial%20Trend%20Analysis_Ransomware%20508%20FINAL.pdf

15 Kasperskycontenthub, "Growth in the Number of Attacks by the Bruteforce. Generic.RDP Family", Accessed 2025-04-02. https://media.kasperskycontenthub.com/wp-content/uploads/sites/43/2020/04/29113731/rdp-stats-all.png

16 Eset, "ESET Threat Report T2 2021 Highlights Aggressive Ransomware Tactics and Intensifying Password-guessing Attacks", Accessed 2025-04-02. https://www.eset.com/int/about/newsroom/press-releases/research/eset-threat-report-t2-2021-highlights-aggressive-ransomware-tactics-and-intensifying-password-guessi

17 Eccouncil, "Cybercriminals Move to Cloud! Hackers Selling Credentials via 'Cloud of Logs'", Accessed 2025-04-02. https://cisomag.eccouncil.org/cybercriminals-move-to-cloud-hackers-selling-credentials-via-cloud-of-logs

18 CNBC, "Leaked Documents Show Notorious Ransomware Group has an HR Department, Performance Reviews and an 'Employee of the Month'", Accessed 2025-04-02. https://www.cnbc.com/2022/04/14/conti-ransomware-leak-shows-group-operates-like-normal-tech-company.html

19 Securitybrief, "What We Can Learn from the Leaked Conti Ransomware Group Chats", Accessed 2025-04-02. https://securitybrief.com.au/story/what-we-can-learn-from-the-leaked-conti-ransomware-group-chats

20 FBI, "Internet Crime Report 2021", Accessed 2025-04-02. https://www.ic3.gov/Media/PDF/AnnualReport/2021_IC3Report.pdf

21 Gatefy, "7 Real and Famous Cases of Ransomware Attacks", Accessed 2022-06-13. https://gatefy.com/blog/real-and-famous-cases-ransomware-attacks

22 Wired, "Atlanta Spent $2.6M to Recover from a $52,000", Accessed 2025-04-02. https://www.wired.com/story/atlanta-spent-26m-recover-from-ransomware-scare

23 Daily-times, "City of Farmington Recovering After SamSam Ransomware Attack", Accessed 2025-04-02. https://www.daily-times.com/story/news/local/farmington/2018/01/18/farmington-recovering-after-ransomware-attack/1044845001

24 Healthcareitnews, "Ransomware Attack on Hancock Health Drives Providers to Pen and Paper", Accessed 2025-04-02. https://www.healthcareitnews.com/news/ransomware-attack-hancock-health-drives-providers-pen-and-paper

25 Csoonline, "Allscripts Recovering from Ransomware Attack that has Kept Key Tools Offline", Accessed 2025-04-02. https://www.csoonline.com/article/3250246/allscripts-recovering-from-ransomware-attack-that-has-kept-key-tools-offline.html

26 Microsoft, "Start Your PC in Safe Mode in Windows 10", Accessed 2025-04-02. https://support.microsoft.com/en-us/windows/start-your-pc-in-safe-mode-in-windows-10-92c27cff-db89-8644-1ce4-b3e5e56fe234

27 Howtogeek, "How to Use System Restore in Windows 7, 8, and 10", Accessed 2025-04-02. https://www.howtogeek.com/howto/windows-vista/using-windows-vista-system-restore

7

Social Engineering Attacks and Mitigation Strategies

Introduction

The digital age has brought about significant changes in nearly every aspect of life, including work, education, shopping, banking, socialization, and entertainment. With the widespread adoption of internet technologies, communication has become faster and more direct than ever before. At the same time, organizations in various industries have turned to digital solutions to streamline their processes and improve communication with customers and partners.

However, the increased reliance on the internet has also made it easier for sensitive information to be shared on social media and other online platforms, often without proper safeguards. Cybercriminals are well aware of this and have developed various tactics, including social engineering (SE) attacks, to exploit publicly available information for their own gain. These attacks seek to manipulate individuals or organizations into divulging sensitive information, such as login credentials, that can be used to gain unauthorized access to IT systems.

While advanced security measures such as firewalls, antimalware, and network detection and response (NDR) have made it more difficult for cybercriminals to target online systems directly, they have turned to exploit the human element, which is considered the weakest element in the cybersecurity chain. As a result, it is important for individuals and organizations to be aware of the potential risks and take steps to protect their sensitive information.

SE attacks are increasing at a rapid pace and have intensified lately. Since the start of the COVID-19 pandemic, most organizations have adopted the work-from-home model to remain operational during the long lockdown period. The massive shift of the workforce to become remote introduced many security problems. For instance, remote employees have to access their corporate resources remotely using their home devices, which are not well-protected comparable with their working environment devices. Cybercriminals have

 DOI: 10.1201/9781003008279-7

ridden the wave and intensified their attacks against remote employees, trying to get an entry point into their corporate networks. SE attacks are considered the main method used by cybercriminals to gain unauthorized access to sensitive IT resources. According to proofpoint[1] report titled "The Human Factor", 99% of cyberattacks utilize SE techniques to deceive unaware users into installing malware.

In this chapter, we will discuss the various types of SE threats that can target users and provide strategies for defending against them. It is important to note that before launching a SE attack, adversaries must first gather information about their targets. This process, known as "Open Source Intelligence (OSINT)", will be covered in the following chapter.

Defining Social Engineering

SE can be defined according to the context where it is used. In the IT security context, SE is a non-technical security threat that utilizes various psychological tricks, through direct connection or via other communication mediums such as phones or the internet, to convince unsuspecting users to disclose sensitive information and to break normal security measures. Cybercriminals normally use SE attacks to steal targets' login credentials to access networks and other protected resources or try to steal their personally identifiable information (PII), such as social security and passport numbers, to impersonate them. Some SE attacks aim to gain intelligence about the type of IT infrastructure and installed security defenses. In contrast, other types try to deceive a user into installing a malicious program to spy on or remotely control the victim's device.

Adversaries find it is much easier to exploit the human tendency to trust to infiltrate target IT systems instead of launching a direct attack against them, such as executing brute-force attacks or exploiting software and hardware vulnerabilities to gain an entry point into the target system.

SE is considered a key element of many cyberattacks. For instance, malware writers use SE tactics to convince unaware users to run malware sent via email attachments. In contrast, phishers use other SE tactics to let unaware users disclose confidential information; scareware operators use SE fear tactics to push the users to purchase useless software programs (e.g., fake antivirus and system maintenance tools) to protect their computing devices – which is completely fake and even harmful (sometimes contains spyware).

SE attacks can be classified according to different criteria; however, we can mainly group them into two groups based on the method used to launch the attack: Internet-based and in-person (physical) attacks.

Before discussing SE attack types, let us discover how the SE attack process works.

The Social Engineering Attack Process

Conducting most SE attacks requires following more than one step. Although each SE attack is unique, most attacks are commonly performed using the following four phases (see Figure 7.1).

Information Gathering

This is the most important phase and is considered the pre-attack phase, as it is executed first before launching the actual attack. First, the attacker needs to identify the key employee within the target organization who owns the sensitive information (e.g., access credentials to key system areas) that can be used to penetrate the target organization's IT system. Attackers try to gather as much information as possible about that specific target and its surrounding environment, so they can later build a trusting relationship with them to improve the chance of a successful attack.

Attackers gather different information about their targets (Both individuals and organizations) in this phase, such as:

1. Contact details include phone numbers, email addresses, and physical addresses.
2. Social media profiles (e.g., Facebook, Instagram, Twitter) to collect personal information and discover the social relations of the target.
3. Target organization geographic address for its headquarters and branch offices, or individual home address.
4. Type of computing devices and IT infrastructure used by the target includes operating system type, IP address, netblocks, installed security solutions (firewalls and IDS/IPS types), networking devices, etc.
5. Technologies used to build the target organization's online presence (website) and applications, in addition to gathering information about its third-party providers (e.g., email system, cloud providers, Managed Service Security Providers (MSSP)).

Social Engineering Lifecycle

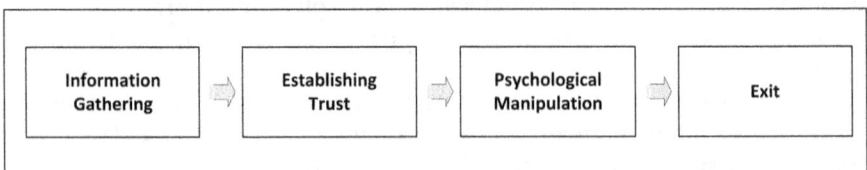

Information Gathering	⇨	Establishing Trust	⇨	Psychological Manipulation	⇨	Exit

FIGURE 7.1
Common SE attack life cycle.

The better information gathered in this step will make the following steps easier.

Establishing Trust

After identifying the possible target to attack, adversaries move to establish a trusting relationship with them. The attacker will use the information gathered in the previous phase to gain the target's trust. For example, the attacker may pretend to know some of the target's friends from college or deceive the target that the attacker is working for a company that is a third-party provider of the target organization.

Psychological Manipulation

After gaining the target's trust, the attacker tries to get as much information as possible from them, such as account credentials or the type of security solutions installed to protect the organization's IT systems.
   This information will be used to:

- Gain unauthorized access to the target organization's network,
- Convince the target to install malware on their device that can be used to collect sensitive information,
- Infect their device – and consequently, target organization – with ransomware or other malware to facilitate launching other attacks, steal sensitive information, or corrupt normal work processes.

The Exit

Finally, after successfully gathering all sensitive information, the social engineer will work to delete any traces that can reveal their identity. For example, the attacker will delete or deactivate all social media profiles that used to talk with the target; if an attacker gains unauthorized access to the target IT system, they will remove the entrance traces. Some attackers plant destructive malware to defect the target system, which hardens the investigation and makes discovering what happened challenging.

Social Engineering Attack Types

SE attacks can be classified using different criteria. The most common one is grouping them based on who conducts the attack (see Figure 7.2) and how it is conducted.

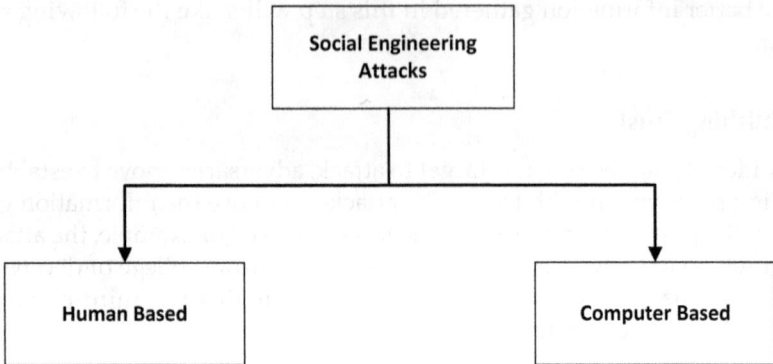

FIGURE 7.2
Social engineering attacks classification.

According to Who Is Involved

Human Based

In a human-based attack, the social engineer will interact directly with the victim to gather sensitive information. In this type, the attack is targeted, and the attacker will use psychological tricks to make the target give the sensitive information using trust or fear. In this way, the attacker can communicate with only one person at a time.

Computer Based

In computer-based attacks, the social engineer will utilize a computing device, such as a computer, laptop, or mobile device (smartphone, tablet), to execute the attack and acquire the sensitive information. Such attacks are mainly carried out over the internet, where the attacker can target a large number of users in a few seconds (e.g., via phishing emails or phishing websites).

According to How It Is Conducted

We can also categorize SE attacks according to how the attack is conducted into the following three groups (see Figure 7.3).

Technical Based

In this type, the attacker utilizes technology – the internet – to gather information about their target using different means. For example, there are many online services for listing stolen users' credentials (username and password) collected from previous data breaches. Suppose a target is using

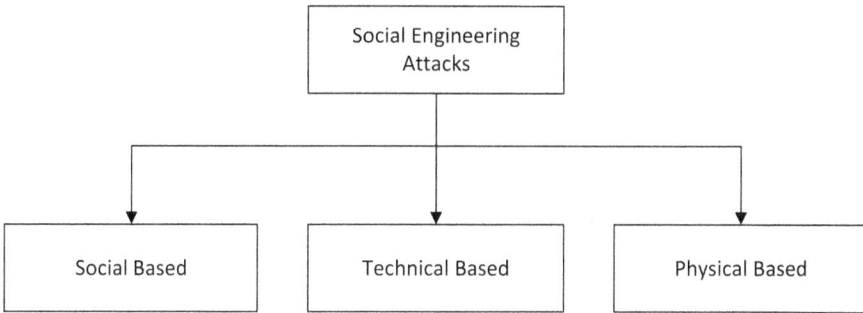

```
                    ┌─────────────────────┐
                    │  Social Engineering │
                    │       Attacks       │
                    └─────────────────────┘
           ┌──────────────────┼──────────────────┐
           ▼                  ▼                  ▼
    ┌──────────────┐   ┌──────────────┐   ┌──────────────┐
    │ Social Based │   │Technical Based│   │Physical Based│
    └──────────────┘   └──────────────┘   └──────────────┘
```

FIGURE 7.3
Classifying social engineering attacks according to how the attack is conducted.

the same password to secure more than one account. Attackers can gain access to all other accounts by knowing one password. Visit the following link for a complete list of websites providing data leak information: https://osint.link/#leak

Data leak websites are not the only service attackers use to find useful information about their targets. A skilled social engineer can utilize search engines, such as Google and Bing, which index a large number of web pages, to find sensitive information about any target. For example, we can use the following Google search query to instruct Google to search any website containing the keyword "John Doe" with titles such as curriculum vitae, phone, or email (see Figure 7.4).

"John Doe" intitle: "curriculum vitae" "phone" "email"

People are increasingly posting personal information to social media platforms such as Facebook, Twitter, and Instagram. Social engineers can utilize these services to gain useful information about their target that can be used later to craft customized attacks against them.

Social Based

This type of attack involves a direct connection with the targets. The perpetrator will use different emotional and psychological tricks to convince the target to provide sensitive information such as credit card info, account credentials, or security questions. Some socially based attacks are also executed over the phone.

Physical Based

As its name implies, the physical-based attack requires the perpetrator to conduct some form of physical action to gather the required information.

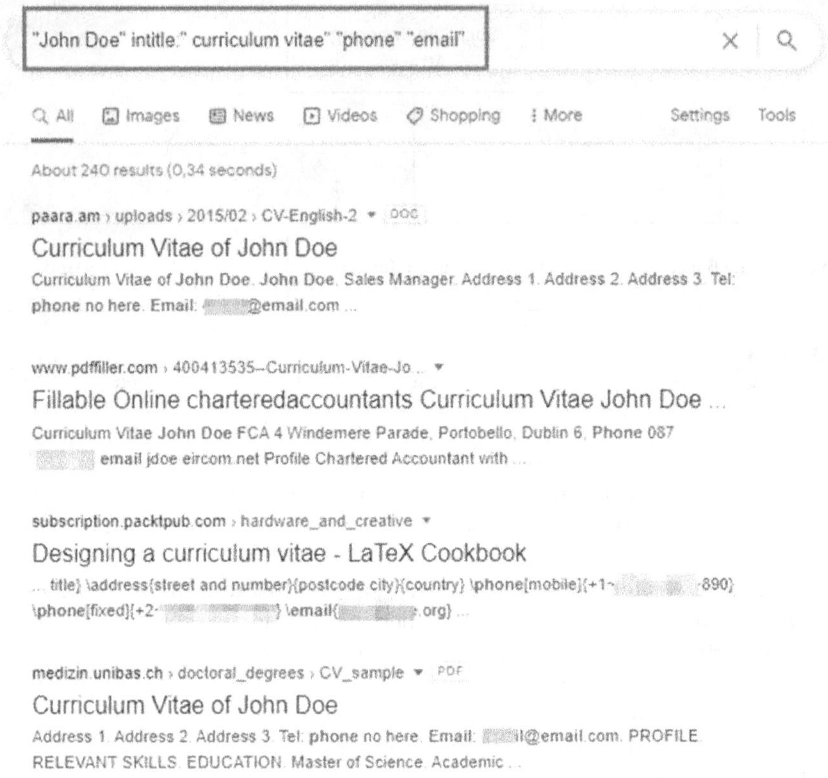

FIGURE 7.4
Utilize Google advanced search query to locate contact information of a specific person.

A popular example is dumpster diving, where attackers search in the trash to find confidential documents and other secret information (such as personal information about employees, suppliers, and other business partners) thrown away without proper destruction.

Social Engineering Attack Techniques

SE can combine more than one attack method to achieve its goals. In this section, we will talk about the most used SE attack techniques.

Phishing

Phishing is the most well-known SE attack. In this attack, the adversary sends an innocent-looking email pretending to be from a legitimate organization, such as a user email service provider, bank, or social media website, and requests the recipient to click a link (which leads to a malicious website) or to reply with sensitive information, such as a user account password. Some phishing emails contain malicious attachments that install malware once the user clicks on them.

The phishing message does not need to be sent via email, although email is the most commonly used medium for sending phishing messages. For example, it could be sent via an instant messaging application (such as WhatsApp), via SMS, or through social media messaging, such as Facebook Messenger.

Phishing messages play on users' emotions to act promptly and without thinking. For example, a typical phishing message asks the user to change their account password immediately to avoid losing access to their account (see Figure 7.5).

Phishing as a category can be further subdivided into other sub-categories.

Spear Phishing

Spearphishing is a customized phishing attack targeted at a specific individual or organization. Adversaries tailor their phishing emails to target specific persons to steal sensitive information or to convince targets to open malicious attachments. While phishing relies on delivering a large volume of spam messages to random users, spearphishing sends one email to one individual who was researched thoroughly before.

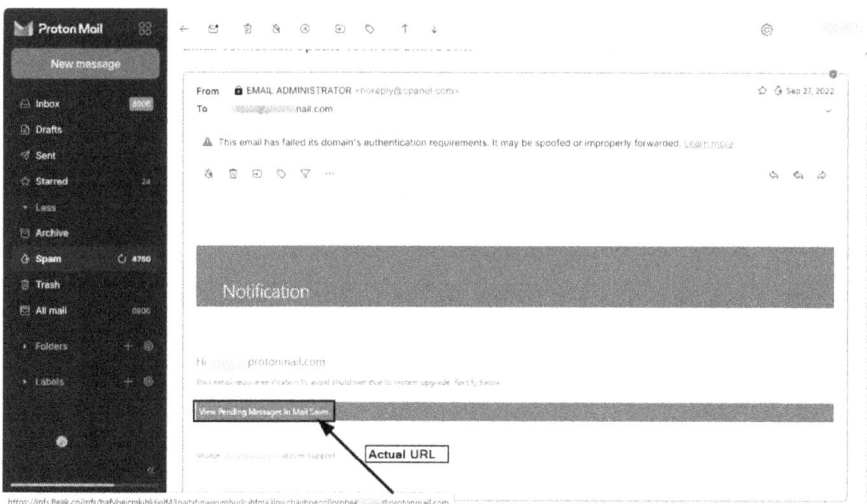

FIGURE 7.5
A phishing email uses fear tactics (to avoid shutdown) to force the user to act promptly. (Verify their email by clicking on the link within the email.)

Adversaries must gather intelligence about their target users before crafting their targeted phishing attack. They commonly use OSINT techniques to collect information from publicly available sources about their targets, such as social media platforms and government databases. A typical spearphishing email will start by addressing the target's name and include important information to draw their attention to continue reading the message and opening the included attachment. The spearphishing email attachment may contain files such as MS Office, ZIP, or PDF documents.

Whaling

Whaling, also known as a whaling phishing attack, is a type of phishing attack that target very important employees within the target organization (and here is why it is called Whaling), such as Chief Executive Officer, Chief Financial Officer, or other corporate staff, to get sensitive personal or business information or to steal their work accounts credentials. Many Whaling attacks try to trick the victims into transferring a large number of funds to the attacker's bank accounts.

Whaling phishing attacks are difficult to detect because it is highly targeted and customized based on the information collected about the target. For example, attackers may create a dedicated website to fool the high-value target into believing it.

Business Email Compromise

Also known as email account compromise (EAC). BEC is a stub-type of phishing attack, and to be precise, it is a subtype of spearphishing attack because it targets a specific person or a group of users within an organization. The main objective of this attack is to fool the victims into conducting unauthorized wire transfers to attackers' accounts or to steal sensitive financial information to perform illegal money transfers. BEC targets all organizations regardless of their size or industry and is growing steadily every year. According to the Federal Bureau of Investigation, the Business Email Compromise cost the world $43 billion of losses between June 2016 and December 2021.[2] The FBI recognizes five types of BEC.

1. CEO Fraud (similar to a Whaling phishing attack): The attackers impersonate the CEO of a company and request fund transfers from an employee in the financial department to a bank account controlled by the attackers.
2. Account Compromise: This attack has become more common due to the increased reliance on cloud technology to facilitate communications with suppliers and other third-party vendors. Attackers hijack one employee account and use it to instruct other employees (e.g., in the finance department) to send wire transfers to vendors while changing the payment details to those belonging to the attackers.

3. False Invoice Scheme: In this attack, adversaries impersonate legitimate suppliers of the target organization and request payments for their services. The scammers will use the same invoice template of the real supplier but change the banking account to the one controlled by them.

4. Attorney Impersonation: As the name implies, in this attack, adversaries impersonate a legal representative, such as a lawyer. This attack commonly targets low-level employees and plays on their fear of obeying attackers' instructions without questioning them for the validity of their requests.

5. Data Theft: In this attack, adversaries try to get sensitive information about an organization's employees from the HR department instead of stealing money. The stolen information can be used to execute other targeted attacks, such as CEO fraud.

Voice Phishing (Vishing)

In vishing attacks, attackers use phone calls to try to obtain sensitive information from their targets. These calls can be made by automated machines or by humans pretending to be from legitimate organizations, such as banks. For example, an attacker may call a target and claim to be from their bank, stating that an urgent issue requires them to update their login information or risk having their account shut down. The attacker hopes to trick the target into revealing sensitive information through this method.

SMS Phishing (Smishing)

In this attack, instead of sending malicious links to target users via email messages, attackers send them via SMS. For example, an SMS message pretending to be from the target person's bank account asking them to click a link to update their banking information to obey the new banking policy; otherwise, they will lose access to their online banking account. However, the malicious link will take the recipient to a website controlled by the attackers and designed to steal their access credentials.

Search Engine Optimization

In this type of attack, attackers create fake websites and use search engine optimization (SEO) techniques, such as keyword stuffing and backlinking, to make them rank highly in search engine results. For example, an attacker may create a fake website that claims to provide information on how to trade cryptocurrency and use SEO to ensure that it appears at the top of search engine results when users search for related keywords. Many internet users assume that websites that rank highly in search results are trustworthy, but in reality, they may be fake sites set up by attackers for malicious purposes. If a user accesses one of these fake websites, the attacker may use an exploit kit to install malware, such as ransomware, on the user's device.

Angler Phishing

Angler phishing is a relatively new phishing targeting social media users. Adversaries impersonate a brand (such as a company or online service) customer service agents on social media websites. They communicate with customers and try to gain their trust to obtain sensitive information, such as personal information or account credentials. An example of Angler phishing is when an unhappy customer posts a negative social media post on Facebook about their experience with their bank (e.g., a customer could be complaining about the delay in executing some banking service). Attackers finding this post can turn it into an angler attack by communicating with the target, pretending to be from their bank, and requesting sensitive personal information to help them with their issue.

In-Session Phishing

In this attack, threat actors interfere during a user browsing session by using pop-up windows. For example, an online banking user may see a pop-up window that appears to be originated from their bank while browsing the bank website. The pop-up window will request the user to provide their account credentials. By doing so, attackers get the user credentials and use them to access their bank accounts and conduct illegal transfers.

Malware

Malware can be used in different scenarios to facilitate or conduct SE attacks. Malware is short for a malicious program; attackers can introduce malware to the target computing device using different methods/techniques, such as via email, visiting a compromised website, downloading internet programs, or attaching an infected USB device. Their final aim is to steal target devices' account credentials and other sensitive information. The most common malware types used in SE attacks are listed below.

Scareware

Cybercriminals use scareware to scare people to visit malicious websites or download and install malicious programs that commonly come in the form of software updates or antivirus and system maintenance tools. Scareware can be introduced to target user devices via pop-up messages while browsing or via phishing email links.

Most scareware attacks aim to scare people into purchasing fake software programs by making them think their device is infected with malware and they need a particular software (which is fake) to fix the issue.

Watering Hole

In a watering hole attack, adversaries compromise a legitimate website with malware; when unaware users visit this website, a malicious script will run silently on their web browser and infect it with malware. This is a targeted attack that targets websites commonly visited by the target organization's employees. For example, a financial discussion forum commonly visited by top management employees of the target organization is a lucrative target for attackers to exploit. Anyone visiting this compromised website may get infected with spaying malware (e.g., backdoor or keylogger) to steal their account credentials and other sensitive information.

Some watering hole attacks could be executed against a software application rather than a website used by the targeted users.

Baiting

In baiting, the attackers provide something useful for the victims to use. For example, a typical baiting attack is throwing malware-infected USB flash drives in the car parking of the targeted organization. When unaware employees pick up these USBs and insert them into their computing devices, they will get infected with malware, such as keyloggers and spyware.

An interesting experiment[3] conducted by the University of Michigan, the University of Illinois, and Google found that 45%–98% of end-users plug in USB drives they find in public places.

Quid Pro Quo

Quid Pro Quo is a type of Baiting attack; however, instead of promising to give the target something of value, the adversary promises to help the target in exchange for some service (a favor for a favor). For example, an attacker could impersonate technical support staff, call a random number in the target company, and ask who asked for a support inquiry. If the attacker identifies an employee who needs some type of support, the attacker instructs them wrongly to gain unauthorized access to their device or request their account credentials to help them fix the issue.

Pretexting

In this attack, the adversary impersonates a legitimate entity, such as a vendor or a banking agent, and tries to obtain sensitive information from the target. For example, the attacker may pretend to be from the target organization's bank, call the finance department, and ask them to verify their banking account information to settle some outstanding money transfers.

Pharming

Pharming is an SE attack where attackers use a technical method to direct unaware users to malicious websites instead of legitimate one. Pharming is achieved technically by manipulating the Domain Name System (DNS) records on the target device to lead internet users to fake websites to steal their credentials or install malware on their devices. A typical pharming attack works as follows.

The attacker sends an email message containing a malicious attachment or a link to a malicious website housing malware. When the target opens the attachment or visits the malicious website, malware (Trojan horse) is installed on their device. The malware will change its HOST file settings. As a result, when the target wants to visit Google.com, it will get directed to a spoofed website instead.

> **Note! What is HOST file?**
>
> The computer HOST file is an operating system text file that maps hostnames to IP addresses. In Windows 10, the hosts file is located at **c:\ Windows\System32\Drivers\etc\hosts**

Attackers may choose to attack the target organization's DNS server instead of end-user devices. In such a scenario, all infected users will have a malware-free device, while the poisoned DNS server will redirect them to the malicious websites instead of the legitimate servers.

Tailgating

This is a type of physical attack. Attackers fool the individual responsible for guarding the secure facility or building of the target company to gain unauthorized access.

Dumpster Diving

Another type of physical attack. In dumpster diving, the attacker searches the organization's trash to retrieve documents containing sensitive information about the target organization and its employees that can help them plan a cyberattack or gain unauthorized access to its IT environment.

You may wonder what sensitive information attackers can find in the trash; here are some examples:

- Employee mail addresses, names, and phone numbers
- Passwords, PINS, and passphrases
- Bank account information
- Copies of official documents
- Audit reports
- Vendors, suppliers, and other third-party providers' names
- Technical guides and instructions that can reveal the type of security solutions and IT infrastructure of the organization

Shoulder Surfing

Shoulder surfing is another SE physical-based attack. In this attack, the attacker steal looks at the target user's computing device screen while typing their access credentials or viewing other sensitive information on the screen to steal sensitive information. Some unaware employees write their passwords on a piece of paper and keep it on their desks. Attackers may try to see this paper to steal the access credentials.

This attack requires the attacker to be very close to the target to succeed.

Social Engineering Prevention

SE attacks rely on psychologically manipulating the human mind to behave against the implemented security controls and the enforced IT security policies. This makes stopping such attacks challenging because people behave differently when faced with an SE attack. For instance, regardless of the implemented access controls and security solutions, attackers can still exploit the "human factor" to penetrate the most protected IT environments.

This section will briefly cover the main measures to protect yourself and your organization from SE attacks. Remember, SE prevention measures involve a mix of security controls, data protection policies, and, finally and most importantly, employee cybersecurity awareness. In the coming part of this book, we will discuss – in more depth – the general security measures to protect endpoint devices and corporate environments from all types of cyberattacks, including SE attacks.

Enforce Multi-factor Authentication

Using an authentication system that still utilizes a password-only system is not secure anymore. Passwords are considered the primary source of cyberattacks. Multi-factor authentication (MFA) requires the user to provide more than one authentication factor to access the protected resources. By implementing MFA, attackers cannot simply gain unauthorized access by acquiring the target username and password combination.

Passwords are vulnerable to cyberattacks, such as brute force, credential stuffing, dictionary attacks, and keyloggers, to name only a few. Enforcing MFA will prevent these attacks and provide multiple layers of security to protect your sensitive resources.

There are different authentication factors – other than traditional passwords – to use when implementing MFA (you should use a combination of these elements); the following list the most common:

- Magic links are sent to the email address
- Hardware and software OTP (one-time password) tokens
- Biometric verification – such as using your fingerprint and Iris scanner to verify your identity
- Authenticator app

Do Not Publish Sensitive Info Online

Before executing their attack, social engineer surfs the internet looking for helpful information about their targets. Nowadays, most people publish a considerable amount of personal and business information on social media platforms. Hackers and other malicious actors can gain insight into a person's interests, activities, and social interactions by monitoring their Facebook, Instagram, and Twitter accounts. For example, sometimes, we can tell if a person is at work or on vacation by checking their Facebook accounts because most people post pictures and make live videos during holidays and post them on social media platforms. Social engineers utilize this information to customize their attacks, such as creating a spearphishing email, which increases the chances that the target person clicks the link within the phishing email.

People should be aware of the type of information they post online. Top management and other employees in sensitive departments, such as IT support and finance, should be trained on SE attacks. They should understand how attackers can exploit any information in their social media posts to infiltrate their online accounts and gain unauthorized access to their organization's IT environment.

Physical Security

Physical security (also known as Physical Social Engineering in this context) prevents access to an organization's physical premises. It deals with the human aspect of physical security. For example, what measures are in place to prevent potential hackers from entering an organization's premises to obtain sensitive information or to access internal systems to conduct malicious tasks – such as:

- stopping online service,
- changing access policy controls,
- making a refund of payments,
- planting malware in target organization's IT environment.

Physical cybersecurity has become increasingly important and is used by hackers. For instance, organizations are now spending large amounts of money on deploying various security solutions to protect their endpoint devices and computer networks; attackers find it easier to attack from an unexpected angle. Going on-site and trying to steal data is more accessible than infiltrating a well-protected network.

Employee Cybersecurity Awareness

Training your employees to respond to potential SE attacks will pay more than installing security solutions. It is essential to train your employees to understand the following:

1. Different types of SE attacks, especially phishing emails.
2. Be very suspicious when receiving communications (by email, phone, or in person) from unknown people.
3. Avoid opening email attachments from unknown senders, and always scan email attachments before opening them using a reputable antimalware program.
4. Avoid giving sensitive information (such as your account password) to unknown people despite an urgent situation.
5. Pay attention to the websites you visit. A common technique hackers use is registering domain names that look similar to legitimate websites they aim to impersonate. For example, in (Figure 7.6), you can see a website pretending to be Grammarly (note how to spell its domain name: https://gnammarly.com) while hackers use it to spread the DarkTortilla malware.
6. Use phishing simulators to see how your employee behaves when faced with an SE attack, especially when receiving a phishing email.

Deploy Various Technical Controls

End-user training is critical to prevent SE attacks; however, this should not eliminate the need for robust technical controls. This section will not delve deep into the types of security controls needed, as we will discuss this in

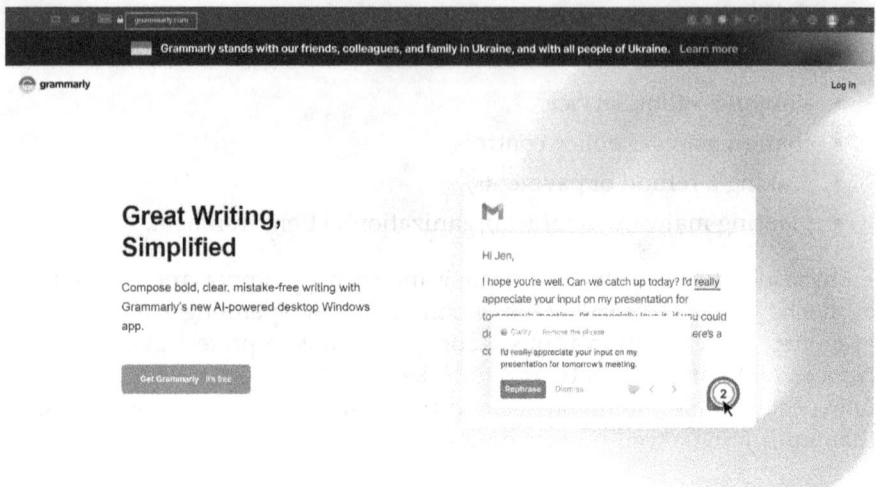

FIGURE 7.6
A malicious website pretending to be Grammarly used to spread the DarkTortilla malware.

more detail in the coming parts of the book (for both individual users and corporate environments). However, for now, remember to implement the following technical controls:

1. Install antimalware programs on network gateways and endpoint devices.
2. Use antispam filters to minimize the volume of spam emails accessing your network.
3. Apply security patches once they become available and keep all installed applications up-to-date.
4. Use an Identity and Access Management (IAM) solution to keep track of all users accessing your IT environment.
5. Use a data classification policy to protect the most sensitive data using the most potent security controls.
6. Make sure to use complex passwords to protect your online accounts. Never use the same password to protect two different accounts, and change your password once every three months.
7. Enforce strict security policies, such as the acceptable usage of company resources, password management, data handling, email, and internet usage using company devices.
8. Organizations should consider using a passwordless authentication system. Passwordless systems, which rely on methods other than

passwords for authentication, can help reduce the effectiveness of many SE attacks. For example, in a passwordless system, an attacker cannot obtain the victim's password through SE techniques such as phishing or pretexting since the victim does not have a password to reveal.

Note: The ISO/IEC 27001 is the world's best-known standard for information security management systems (ISMS) and their requirements. Provide gaudiness to protect your information in all shapes: paper-based, cloud-based, and digital data. You can check its homepage at: https://www.iso.org/isoiec-27001-information-security.html

Summary

SE attacks work by psychologically manipulating people's minds to gain sensitive information, such as access credentials. A social engineer uses urgency to make the unaware user feel fear and act quickly without thinking about the result.

SE attacks can be executed online, via phone calls, or in-person. The most used method is email communications to lead unsuspecting users to click a malicious link or download a malicious attachment containing stealer malware. Most SE attacks begin by gathering information about the target to customize the attack accordingly. Attackers leverage OSINT methods and techniques to collect information about their targets using publicly available information, such as information available on social media platforms, government databases, and search engines. For example, an attacker could use information collected through OSINT to craft a personalized phishing email or to impersonate someone online to gain access to sensitive information or protected resources. It is essential for individuals and organizations to be aware of the potential risks of OSINT and to take steps to protect themselves against these types of attacks.

In the next chapter of this part, we will discuss OSINT methods and techniques. OSINT refers to the collection and analysis of publicly available information from various sources, including the internet, social media, and traditional media. Businesses use OSINT in multiple ways, such as competitive intelligence, customer intelligence, market intelligence, social media intelligence, and fraud investigation. However, it is critical to understand how cybercriminals leverage OSINT to execute attacks against individuals and organizations, which we will cover next.

Notes

1 ProofPoint, "The Human Factor 2022", Accessed 2025-04-02. https://www.proof-point.com/us/resources/threat-reports/human-factor
2 IC3, "Business Email Compromise: The $43 Billion Scam", Accessed 2025-04-02. https://www.ic3.gov/PSA/2022/PSA220504
3 Googleusercontent, "Users Really Do Plug in USB Drives They Find", Accessed 2025-04-02. https://static.googleusercontent.com/media/research.google.com/en//pubs/archive/45597.pdf

8

Open Source Intelligence Techniques (OSINT)

Introduction

In the previous chapter (Social Engineering Attacks), we saw that in a targeted SE attack, the social engineer needs to gather information about their targets before crafting a customized attack. For example, in Spearphishing attacks, adversaries need to tailor email content to make the recipient trust the sender to visit a malicious website or download and open a malicious attachment. However, did you think about where hackers collect this information about their targets?

Threat actors search all publicly available information, such as social media platforms, government databases, and corporate websites, to name only a few, to find any information that can help them in their attack. They may also use specialized tools such as Wireshark and vulnerability scanners to discover vulnerabilities in target computer networks to exploit.

On the other hand, ethical hackers and penetration testers use the same search techniques to discover sensitive data and vulnerable systems. The early phase of any pen-testing methodology begins with information gathering.

The Penetration Testing Execution Standard (PTES) (http://www.pentest-standard.org/index.php/Main_Page), which is my preferred pent-testing methodology, consists of seven phases. Intelligence gathering is the second phase after identifying the engagement scope. The PTES defines the Intelligence Gathering phase as *"performing reconnaissance against a target to gather as much information as possible to be utilized when penetrating the target during the vulnerability assessment and exploitation phases. The more information you are able to gather during this phase, the more vectors of attack you may be able to use in the future."*[1]

There are various ways to collect information from public sources. The widespread usage of internet technologies and the massive number of users worldwide make large volumes of personal and business information

DOI: 10.1201/9781003008279-8

available online. You can get a fair amount of details about anyone world-wide by checking their social media accounts and conducting a deep search using government databases.

The act of collecting information from publicly available sources – mainly the internet – and analyzing it to create actionable intelligence is called open source intelligence (OSINT).

This chapter will cover the concept of OSINT, see how it can be used in different scenarios to produce valuable intelligence, and understand the different user groups who leverage OSINT in their work. We will also talk about different tools and online services to gather OSINT, give a practical example of using OSINT to investigate the ownership of a website, and end the chapter by discussing the legal aspects of OSINT gathering. However, before we start, let us define OSINT in detail.

Further Reading:

It is impossible to cover all aspects of OSINT in just one chapter. To gain a thorough understanding of the topic, it is recommended to read a dedicated book on OSINT gathering, such as "Open Source Intelligence Methods and Tools: A Practical Guide to Online Intelligence", 1st ed., authored by Nihad A. Hassan and published by Apress in 2019.

Defining Open Source Intelligence

OSINT refers to the practice of collecting information from publicly available sources. This includes searching online and offline sources; however, all information should be legally obtained without breaching copyright or privacy laws. OSINT practitioners use advanced search techniques to find information in a massive, visible data volume. Then, they use their expertise to analyze the gathered information and connect the intersections to produce actionable intelligence results.

OSINT goes beyond what can be found using regular search engines like Google and Yahoo. While the information on the surface web is helpful for OSINT, the majority of web data is located on the deep and darknet, which cannot be accessed through regular search engines.

Historically, OSINT emerged as an affordable source in addition to covert spying activities, such as phone tapping, mail monitoring, and reconnaissance satellites. Government agencies found it helpful to monitor publicly available information, such as newspapers, TV, and radio broadcasts, to gather intelligence. This occurred well before the advent of the internet and social media. The term OSINT was coined to describe this type of intelligence gathering.

Note:

OSINT differs from the open-source software term used to describe creating open-source code software.

With the increasing use of technology and the growth of online interactions, OSINT has become increasingly popular among cybersecurity professionals and hackers. Ethical hackers use open-source intelligence gathering to identify various entry points into an organization's IT environment, including physical, digital, and human-based vulnerabilities. For example, employees may not be aware of the amount of personal information available online that cybercriminals could exploit to craft a targeted attack. On the other hand, organizations may not be aware of the amount of sensitive information publicly available online that could be used against them by threat actors.

OSINT Sources Types

OSINT can be sourced from both offline and online sources, but with the widespread use of internet technology, most sources are now found in the digital realm. Here is the main source of OSINT.

Internet

Booted by the huge adoption of technology worldwide, the internet is considered the main source of OSINT. Internet sources include everything that can be found online, such as:

Blogs, forums, job websites, search engines, government databases (e.g., Vital, courts, and property records), corporate records, data leak websites (such as Pastebin sites), social media platforms (e.g., Facebook, Twitter, Mastodon), WHOIS information, website history and anything that can be found online and is publicly accessible.

Traditional Media

TV, radio broadcasts, newspapers, magazines, road advertisements, paper books, films, art, public notice, flyers, and other forms of print.

It is worth noting that many forms of traditional media still exist. However, they are now shifted to become fully or partly digital, such as newspapers, magazines, and books. On the other hand, many TV and radio stations have also added online broadcasting options.

Grey Literature

Grey literature includes all materials published outside the normal publishing and media distribution channels. Government agencies, non-government organizations, private consultants, commercial organizations, and academic centers commonly produce it. The most common grey literature publication types include:

- Reports from charities and other non-profit organizations
- Academic journals, research papers, dissertations, scientific papers,
- Annual business reports and official corporate documents – such as company profiles, company news, annual meetings, whitepapers, and filing data
- Leaked corporate records. A good example is the leaked documents of PANDORA PAPERS[2], which exposed a shadow financial system that benefits the world's richest and most powerful people.

Grey literature materials could be distributed internally within organizations or through specific channels. Most channels require paying a subscription fee to access grey materials.

> Note:
> A comprehensive list of document types in Grey Literature can be found at: http://www.greynet.org/greysourceindex/documenttypes.html

Metadata Information

Metadata is information that describes other information. In the digital world, many different types of files, such as videos, pictures, MS documents, PDF files, and more. Each type of file has its own structure and some information about who made it, the date/time when it was made, and other details. Some files, like pictures and videos, might have information added by the person who made them. Other files might have information added by the device or program that created them, like the type of camera used to take a picture or the photo editing program used to edit or create a drawing.

Geospatial Information

This includes maps, satellite imagery, and location-based social media data. Some digital file types, such as images, can contain GPS data indicating the geographical location where the photo was taken.

Intelligence Collection Disciplines

There are primarily five types of intelligence sources. For this book, we are concerned about cybersecurity; hence, we will focus on OSINT gathering.

Human Intelligence (HUMINT)

HUMINT gathers intelligence from human sources and directly communicates with people (e.g., civilians, refugees, local inhabitants) to acquire the required information.

Communication with the target people can be conducted either by directly engaging with them or through spies (or foreign agents) trying to collect information from human sources through conversation or interrogation.

Signals Intelligence (SIGINT)

SIGINT refers to the act of capturing electronic communications through ships, satellites, planes, and spy ground stations to acquire intelligence. Communications Intelligence (COMINT) is a subtype of SIGINT and refers to intercepting communications between two individuals. The intercepted communications could be conducted via SMS, phone calls, and online communications (such as email, internet messaging, and private social media messages – for example, Facebook messengers).

The National Security Agency (NSA) conducts SIGINT in the USA.

Imagery Intelligence (IMINT)

IMINT is the practice of analyzing and constructing useful intelligence (e.g., converting electronic data into visual displays or graphics) from data collected via spaying satellites, aerial photography, or radar.

Measurement and Signatures Intelligence (MASINT)

MASINT is a little-known intelligence-gathering schema concerned with collecting information about weapons capabilities and industrial activities related to developing weapons systems.

MASINT leverages data collected from IMINT and SIGINT gathering systems.

Open-Source Intelligence (OSINT)

OSINT includes all information that can be accessed publicly in print or electronic form without violating any data privacy laws. With the advancement of internet technologies, online sources have become the primary source of OSINT gathering.

Intelligence Gathering Types

We can distinguish between three methods when gathering intelligence.

Passive Information Gathering

This is the essence of OSINT activity. Gathering intelligence from public sources without alerting or notifying the source about the gathering activity. For instance, browsing the target company's website, reading articles and news about a particular corporation, collecting contact information from the target corporation's website, checking the previous history of a target website via the Wayback Machine service, and monitoring social media posts are all examples of passive information gathering.

Active Information Gathering

We interact directly with the target IT infrastructure or person in this information-gathering type. This interaction will be detected on the target end and could be discovered as a malicious attempt to attack the IT systems or the individual target device. There are different examples of active information gathering, such as:

- Mapping target network infrastructure using Nmap
- Pent-testing target IT infrastructure to discover weak entry points
- Conducting different social engineering attacks (e.g., phishing, baiting, and pretexting) to acquire sensitive information
- Using specialized tools for vulnerability scanning
- Interacting personally with the target individual to acquire intelligence

Semi-passive Information Gathering

This information-gathering type falls between passive and active information gathering. It involves interacting with the source slightly without leaving a trace or alerting it about the gathering activity. Here are two examples of using semi-passive information gathering.

- Using a search engine, such as Google, to find information about a particular individual or company. When searching for someone using Google, the target will not get notified about this search.
- Sending a small packet to the target IP address or website that resembles normal internet traffic to detect open ports and running services.

Semi-passive will not leave a clear digital footprint about our information-gathering endeavors.

OSINT Users and Example Use Cases

OSINT can be used by a wide range of organizations and individuals for varying purposes. In this section, we will talk about the primary users of OSINT.

Government

Government agencies are among the top users of OSINT gathering. For instance, security intelligence services and law enforcement agencies use OSINT to investigate crimes, collect information about suspected people, and prevent crimes before they occur. OSINT is also leveraged to counter terrorism by monitoring social media platforms and communications taking place in the darknet between threat actors.

In the cybersecurity field, law enforcement uses OSINT to protect national public infrastructures from attacks originating from cyberspace. Such as attacks against:

- The internet and telecommunications services
- Electricity grids and water supply infrastructure
- Transportation infrastructure such as airports, main highways, and seaports

OSINT is also helpful for government departments other than security, as it provides insight into trending issues in local societies and predicts chaos and potential unrest events before they occur or escalate. In such scenarios, OSINT is considered a valuable source of intelligence for policymakers to help them make better-informed decisions.

Corporations

OSINT has become an indispensable tool for business corporations. For instance, OSINT can be leveraged in this field in the following ways:

- Cyber Threat Intelligence: Just like government agencies, corporations can gather valuable threat information by monitoring OSINT sources on the surface web and dark Web. For example, if a company experiences a data breach, cybercriminals may advertise the stolen employee credentials on dark web marketplaces to sell to other criminal groups. Another way OSINT can be used in cyber threat intelligence is by monitoring dark web chat rooms and discussion forums where threat actors may plan attacks against a specific organization. By being aware of this information in advance, businesses can take steps to protect their assets and prevent attacks by cybercriminals.

- Competitor intelligence: OSINT can be leveraged to monitor other competitors' news and marketing strategies. For example, by monitoring competitors' social media accounts, we can get good information about their future product/service launches and customers' opinions about these products or services.

- Market intelligence: OSINT can be used to understand market trends in a specific community or market. What people like or dislike and their needs in a specific market can be discovered by monitoring samples of users' social media accounts and online public interactions in a particular community or market.

- Fraud detection: OSINT is used to connect people with companies and determine their reliability. Researching companies can also reveal important information about their work histories and business partnerships. For instance, some companies may change their business name over time to mask their previous suspicious business relationships.

- Brand protection: Companies use OSINT to search for their products and discover if someone is counterfeiting them.

- Brand reputation: Businesses use OSINT to keep an eye on social media posts and other news sources that mention their brand name to counteract any negative or false reviews. This helps them defend their reputation and ensure that accurate information about their company is disseminated.

- Background checks: Search publicly available sources to gather information about a specific person or firm's background to determine their trustworthiness in doing business with them.

Ethical Hackers

Ethical hackers and pen-testers use OSINT techniques to gather intelligence in the same way threat actors do to:

- Discover entry points into the target organization's IT environment
- Discover leaked sensitive information about the organization published on the darknet forums
- Stimulate conducting social engineering (SE) attacks to help train and educate employees and end users about SE threats.

Cybercriminals

Most cyberattacks, particularly those targeted and sophisticated, such as ransomware and APT attacks, begin with the phase of gathering information. Hackers use OSINT to gather a plethora of information about their intended targets, whether it's a company or an individual, before crafting a personalized attack. This information-gathering phase is crucial as it enables threat actors to understand the target's vulnerabilities and devise an attack strategy that is more likely to be successful.

Privacy-Conscious Persons

Individuals can use OSINT techniques to uncover sensitive or personal information they may have inadvertently shared online. Many people have become aware of the vast amount of personal information collected by large technology companies, such as social media platforms and internet service providers. Using OSINT techniques helps them discover such information and delete it to ensure their privacy. Furthermore, OSINT can assist individuals in identifying identity theft attempts. For instance, an adversary may try to impersonate someone by creating social media accounts using their name and personal photo. OSINT can detect such attempts and help maintain internet users' security and privacy.

OSINT Advantages

As we saw in the previous section, OSINT can be leveraged by different user groups to support their needs in different scenarios. There are numerous benefits that make OSINT stand apart compared with other intelligence domains. The following list the main ones:

- Cost-effective: OSINT relies on gathering intelligence from publicly available sources. Compared to traditional information-gathering types, it does not require heavy resources to perform it. Organizations with limited intelligence budgets will find OSINT an indispensable tool to acquire high-value intelligence for a low cost. For instance, a computer and an internet connection are all that you need to start your gathering activity.
- Access to real-time information: OSINT allows intelligence from social media platforms to be gathered in real time. This helps decision-makers make informed decisions about urgent cases instantly.
- Access to wide resources: OSINT sources are spread across online and offline resources. Beginning with commercial databases and

social media platforms and ending with government databases. The OSINT researcher will have access to a wide range of resources, making their work more comprehensive.

- Sharing information legally: OSINT information can be shared freely with anyone without the fear of breaching any compliance or data privacy regulation worldwide because it is legally obtained from publicly available sources.
- Ensure compliance with rules: OSINT can be leveraged to ensure a business is confirming compliance with rules and other enforced data privacy regulations.
- Variety of resources: OSINT resources come in different forms, such as text, image, video, and audio. Investigating the metadata within these file types can also reveal important information.

OSINT Techniques

When searching for OSINT resources, you must use different techniques to gather information. In this section, we will list the most important gathering techniques.

Search Engines

Search engines locate information from the huge volume of resources available online. When thinking about search engines, Google is the first one that comes to mind. However, there are many more. For instance, search engines can be broadly categorized according to their functionality into:

- General-purpose search engines (also known as Mainstream search engines): These can be used to search surface web content. They use a crawler or spider to index web content and store the results in a huge database that can be searched by users using keywords. Mainstream search engines track users when conducting searches through them to display targeted advertisements. Examples of this type are Google, Bing, and Yahoo!
- Privacy-oriented search engines: These are similar to mainstream search engines; however, they do not collect users' personal information when interacting with their service or track their web browsing activity across the Web. Examples of such search engines are: DuckDuckGo and Startpage.
- Specialized search engines: This type searches specific niches and does not search the entire web content. Examples of this type are

Amazon and eBay's internal search functionality, search features of social media platforms such as Twitter and Facebook, and the Google Scholar search engine.

- National search engines: These are used to search within the content or websites of a specific country. For example, Baidu (https://www.baidu.com) focuses on searching within Chinese websites.
- Metasearch engines: These are used to search multiple search engines at once and aggregate the results back to the searcher in a single display. For example, Dogpile.com returns results from both Google and Yahoo! search engines.

Other types of search engines are used to search for specific niches, such as:

- People search engines – such as: peoplefinder.com and spokeo.com
- Corporate & business search engines – such as: opencorporates.com
- Phone numbers search engines – such as: truecaller.com
- Files search engines like faganfinder.com and heystacks.com for searching public Google documents.
- Reverse image search engines: Mainstream search engines such as Google, Bing, Yahoo!, and Yandex allow searching for images.

It is necessary to remember that conventional search engines will search for content available on the surface web. Contents buried in deep and the dark nets cannot be indexed, so typical search engines cannot discover them.

Note!

You can use advanced search operators in Google to narrow down your search results. For example: **"search query" filetype:pdf** will search for all PDF files with the keyword **"search query"**.

In the same way, if you want to search for "search query" in all PDF files within a particular website, type the following in Google:

"search query" filetype:pdf site:domain.com

Social Media Intelligence (SOCMINT)

In today's digital age, it is rare to meet someone who has an internet connection without owning at least one social media account. According to Investorbrandnetwork[3], there will be more than 4.41 billion social media users in 2025.

Social media intelligence is considered a sub-branch of OSINT; it is concerned with acquiring information from social media platforms, such as Facebook, Instagram, Twitter, and LinkedIn, and using it in our investigations.

Social media platforms are not limited to popular social networking websites, such as Facebook and LinkedIn. For instance, discussion forums, blogs, social gaming, social bookmarking, and microblogging platforms like Mastodon and X, in addition to photo and video-sharing websites, all fall under this category.

You can get a great deal of information about anyone worldwide by checking their profiles on different social media platforms. Users' job history, friends, social interactions, political and religious views, personal photos, videos, and current locations can be acquired by monitoring their social media profiles.

OSINT leverages social media in multiple ways by different actors:

- Background check about individuals.
- Monitoring organization's social media presence to acquire useful information about their activity, monitoring customers' interactions with the brand, and comparing one organization's work with its competitors.
- Profiling people on a large scale by national security agencies.
- Predicating events worldwide through monitoring the social media profiles of people living in a specific country or society to predict events such as unrest, political instability, and disaster prediction and response.

Different types of information can be gathered from social media platforms, such as:

- User profile information – For example, on LinkedIn, you can find current and previous work locations. On Facebook, you can know a person's spouse and social interactions with friends and relatives.
- Digital files – Such as user-posted photos, videos, and text posts.
- Metadata – This includes information hidden within a user file, geolocation information, and the date/time when a particular piece of information was made.

It is worth noting that some OSINT investigations consider information acquired from social media platforms to be private information because it was intended to be used on these platforms only and shared with specific people. Investigators sometimes need to use fake accounts to access other people's profiles on social media platforms, and this could be considered against the law in some cases and against the "usage terms" of these platforms.

On the other hand, many investigations consider SOCMINT to be entirely legal because information published on them was meant to be published publicly, making it belong to OSINT resources.

Data Mining

OSINT can result in large amounts of data during gathering. To analyze this data, data mining techniques can be applied to find patterns, trends, and connections. These techniques are used by government agencies to collect information from various online sources, including social media, news sources, and blogs. Data mining helps categorize this information and make predictions about future events, such as economic instability in a country, based on social media posts and economic news collected from public news outlets.

Web Scraping

Web scrapers are tools used to extract information from websites. OSINT investigators may need to harvest particular information from a website automatically, which requires collecting unstructured information from the target website's HTML source code and processing it in a structured and clean format (such as storing it in an MS Excel file) to be used in the investigation.

A good example of using web scraping in OSINT is to use it to collect email addresses and contact information of all users presented on a particular website. Web scraper can be instructed to collect other information, such as inbounds and outbounds URLs available in a website to discover possible connections with other websites.

There are different types of web scrapers; some are ready to use out of the box (pre-built), while others are self-built from scratch. The self-built scrapers can be easily customized to fit the data collection requirements of a specific website. Web scrapers can collect information about a specific entity or individual from social media platforms (e.g., a Facebook friends list, an X account follower, and others).

Note!

Python is commonly used to develop web scraping scripts. **Scrapy** (https://scrapy.org) is a famous open-source crawling framework developed using Python. **Beautiful Soup** (https://pypi.org/project/beautifulsoup4) is a Python programming library for web crawling.

OSINT Tools

When searching for information, OSINTT gatherers need to use different tools and online services to help them find the required information. In this section, we will list the primary OSINT tools that should exist with any OSINT researcher.

Advanced Google Dorks

We already talked about using search engines to find information on the publicly accessible part of the internet. However, some search queries can return millions of results, and OSINT researchers need the most relevant results to reduce analysis time. Google allows the use of special operators that help modify search queries to return the most relevant results. For example, the following Google dork:

> inurl:osint filetype:docx | filetype:doc | filetype:odt | filetype:rtf | filetype:pdf | filetype:txt

will search for websites that contain the string "OSINT" in their URL and limit the results to files of types pdf, docx, doc, odt, rtf, and txt. There is a website called the "Google Hacking Database", that lists hundreds of advanced search queries to help OSINT researchers return specific results. The website can be found at https://www.exploit-db.com/google-hacking-database.

Whois

The Whois database contains a list of all registered domain names worldwide, their owner name, and contact information. By checking the Whois records, we can reveal the ownership of websites and IP addresses. For instance, we can get various information about any domain name by checking its Whois record, such as:

- Website admin name, email, and other contact information such as phone number and mailing address
- Domain name history
- Domain name creation and expiration date
- Information on where the domain name is hosted
- Contact information of the domain registrar (the provider or corporation that registers the domain name)

The information included in the Whois database can be very useful for OSINT investigations. For instance, after getting a specific domain owner's email address, we can conduct an email search to reveal all places where this email was used – such as leaked databases, public posts, or comments including this email address, and find all domain names registered using the same email address.

> **Note!**
> Some website owner hides their personal information in the Whois database by paying a small fee. When viewing the Whois information of such domain names, it will display another company information (the proxy company), not the real owner's personal information.

Urlscan.Io (https://urlscan.io)

Urlscan is a free online service that analyzes websites. When submitting a specific domain name for Urlscan, it will browse the website similar to a regular user and monitor all connections that a web page has made, such as:

- The domains and IPs contacted
- The resources contacted upon loading the web page, such as JavaScript, CSS, images, audio, videos, and other file types
- Cookies created by the web page
- Whois analysis of the submitted domain name
- Sub-domain names
- IP address information
- Technologies used to build the website, such as WordPress, Joomla, and used programming libraries, such as JavaScript (JQuery)
- JavaScript global variables – this can be helpful to identify client-side frameworks and code
- Links to other websites
- The Document Object Model (DOM) tree of the web page
- The API requests made by that page

Wayback Machine (https://archive.org/web)

The Wayback Machine helps us reveal important information about previous versions of websites and other online contents. It contains an archive of the Web and is considered an indispensable tool for OSINT gatherers to reveal important information stored in the historical versions of websites, such as deleted content, hidden files, updated content, old images, and videos. For example, when inspecting an archived version or screen captures of a particular website, we may find contact details and other information that reveal important communications with other entities.

The Wayback Machine started archiving web content beginning in 1996. There are billions of web pages (783 billion web pages at the time of writing) stored in its database, and it continues to grow daily.

Shodan (https://www.shodan.io)

Shodan (Sentient Hyper-Optimized Data Access Network) is the first search engine specialized in searching for all internet-connected devices. It allows you to search and find all internet-capable devices, such as power plants, mobile phones, servers, routers, webcammers, home appliances, smart fire alarms, smart door locks, smart bicycles, medical sensors, and any machine or device accessible via the internet.

According to Statista, the number of Internet of Things (IoT) devices worldwide is expected to reach 29 billion IoT devices in 2030.[4] Shodan provides a way to discover and gather information about internet-facing devices, such as all devices connecting to the internet from a specific network.

Many use cases for Shodan can be leveraged in OSINT. For example, we can use it to locate information about a specific IP address, such as:

- Identify the device connecting via this IP
- Open ports and running services on the IP
- Historic IP data, such as how long the device has been online
- Search for specific devices in a particular subnet
- Discover netblocks associated with this IP address and discover more IP addresses that belong to the same network or organization to investigate.

Source Code Search

In the OSINT context, code search is concerned with searching for programming-related content, such as code snippets and scripts, in public repositories, such as developers' forums and code repositories. Code search can be used to locate connected websites, search for where a specific vulnerable code is used, and to identify software dependencies of a particular application.

There are dedicated search engines for finding source code; the following are the most popular:

- Google Code search (https://developers.google.com/code-search)
- GitHub Search (https://github.com/search)
- GitLab Search (https://docs.gitlab.com/ee/user/search/advanced_search.html)
- Search Code (https://searchcode.com)
- Android Code Search (https://cs.android.com)

theHarvester (https://github.com/laramies/theHarvester)

theHarvester is a free command-line tool developed using Python programming language for conducting reconnaissance. It comes preinstalled with Kali Linux. The tool can collect various information such as names, emails, IPs, subdomains, open ports/banners, and URLs by using various public sources (e.g., search engines, PGP key servers, and SHODAN).

Phone Number Research

If you have a phone number and you want to discover who is behind it, you can use a service such as Truecaller (https://www.truecaller.com) OR NumLookup (for USA only https://www.numlookup.com) to locate the owner name of the phone number.

There are many more online services for conducting reverse phone searches; some are connected with public records and display various personal and work-related information about the owner of the phone number. However, reliable services require user registration and payment to provide accurate results.

hunter.io (www.hunter.io)

This is a tool for finding all emails associated with a specific domain name. It will display the source (the URL) where each email address was found.

Have I Been Pwned (https://haveibeenpwned.com)

This is an online service for finding if a particular email address has been involved in a data breach.

ExifTool (https://exiftool.sourceforge.net)

ExifTool by Phil Harvey is a popular open-source tool for extracting metadata (see Figure 8.1) from different digital file types, especially pictures and video files. ExifTool supports many different metadata formats, including EXIF, GPS, IPTC, XMP, JFIF, GeoTIFF, ICC Profile, Photoshop IRB, FlashPix, AFCP and ID3, Lyrics3.

Linking Different Social Media Accounts Together

As we know, most internet users have one or more accounts on different social media platforms. According to Demandsage,[5] the average number of social media accounts held by millennial or Gen Z users around the world is 8.5. Many people prefer to use the same handle (or username) on all their social media accounts. Suppose we know the social media handle of a particular user on Facebook, and we can use this info to find all accounts on other social media websites that use the same handle.

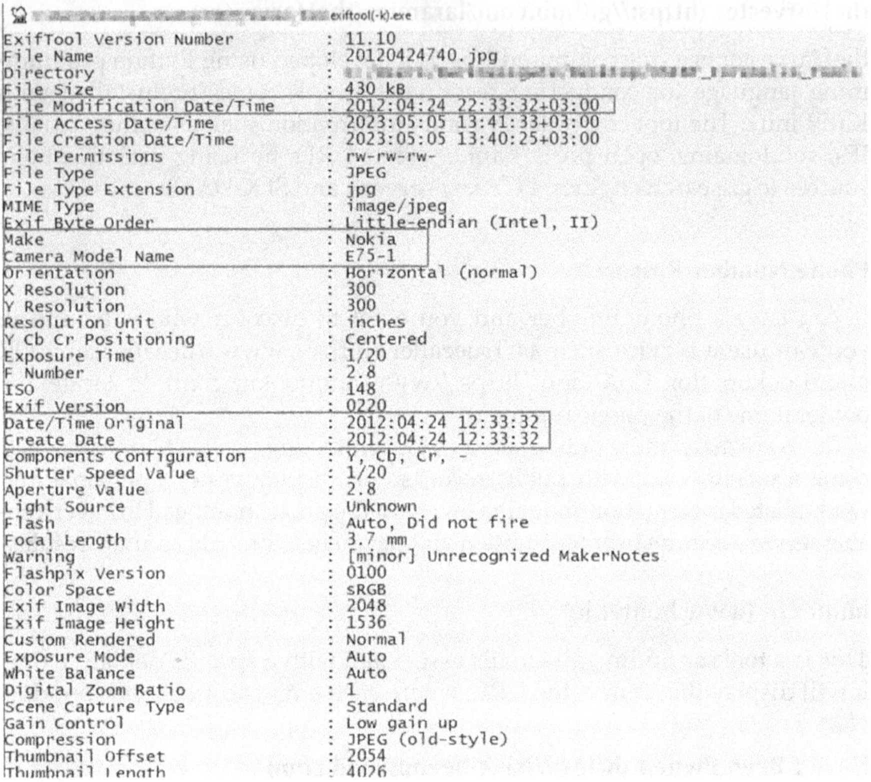

```
                                                     exiftool(-k).exe
ExifTool Version Number     :  11.10
File Name                   :  20120424740.jpg
Directory                   :
File Size                   :  430 kB
File Modification Date/Time :  2012:04:24 22:33:32+03:00
File Access Date/Time       :  2023:05:05 13:41:33+03:00
File Creation Date/Time     :  2023:05:05 13:40:25+03:00
File Permissions            :  rw-rw-rw-
File Type                   :  JPEG
File Type Extension         :  jpg
MIME Type                   :  image/jpeg
Exif Byte Order             :  Little-endian (Intel, II)
Make                        :  Nokia
Camera Model Name           :  E75-1
Orientation                 :  Horizontal (normal)
X Resolution                :  300
Y Resolution                :  300
Resolution Unit             :  inches
Y Cb Cr Positioning         :  Centered
Exposure Time               :  1/20
F Number                    :  2.8
ISO                         :  148
Exif Version                :  0220
Date/Time Original          :  2012:04:24 12:33:32
Create Date                 :  2012:04:24 12:33:32
Components Configuration    :  Y, Cb, Cr, -
Shutter Speed Value         :  1/20
Aperture Value              :  2.8
Light Source                :  Unknown
Flash                       :  Auto, Did not fire
Focal Length                :  3.7 mm
Warning                     :  [minor] Unrecognized MakerNotes
Flashpix Version            :  0100
Color Space                 :  sRGB
Exif Image Width            :  2048
Exif Image Height           :  1536
Custom Rendered             :  Normal
Exposure Mode               :  Auto
White Balance               :  Auto
Digital Zoom Ratio          :  1
Scene Capture Type          :  Standard
Gain Control                :  Low gain up
Compression                 :  JPEG (old-style)
Thumbnail Offset            :  2054
Thumbnail Length            :  4026
```

FIGURE 8.1
Using ExifTool to extract metadata information from a JPG image file.

Manual search for a user handle on different social media accounts is a daunting task if we want to make it manual. However, by using dedicated tools, we can speed up the process. Here are some services to conduct reverse username searches.

- Namecheckup (https://namecheckup.com)
- Sherlock (https://github.com/sherlock-project/sherlock) – a command-line tool for searching for a particular username across a large number of social media platforms (about 406 sites at the time of writing) (see Figure 8.2).

Geospatial Research and Mapping Tools

Geospatial research is concerned in analyzing geospatial data to find physical locations on Earth. For example, a GPS coordination could be found within an image EXIF metadata; we can use mapping applications such as Google Maps to locate that location on Earth. On other hand, we may need to

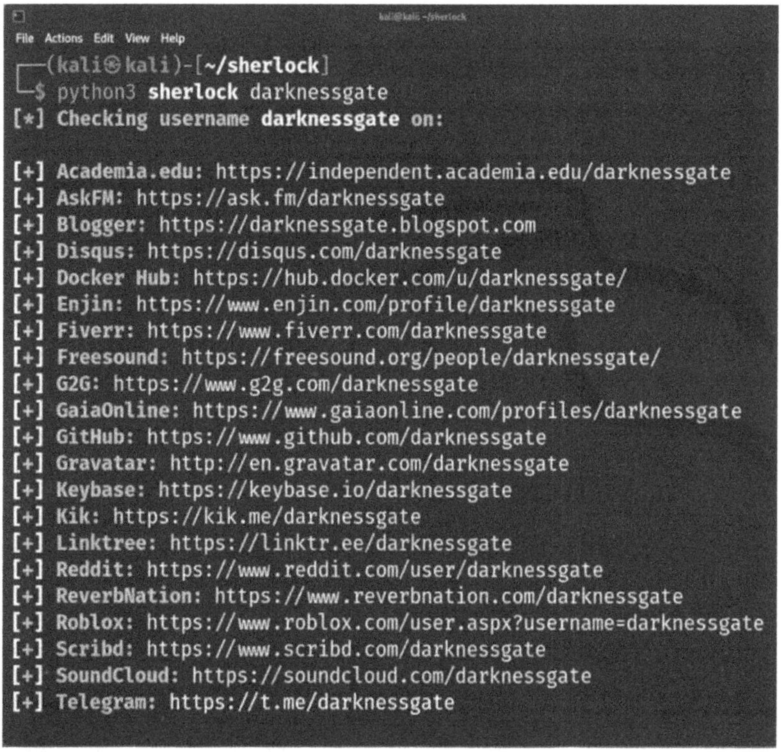

```
File Actions Edit View Help                    kali@kali ~/sherlock
  ┌─(kali⊛kali)-[~/sherlock]
  └─$ python3 sherlock darknessgate
[*] Checking username darknessgate on:

[+] Academia.edu: https://independent.academia.edu/darknessgate
[+] AskFM: https://ask.fm/darknessgate
[+] Blogger: https://darknessgate.blogspot.com
[+] Disqus: https://disqus.com/darknessgate
[+] Docker Hub: https://hub.docker.com/u/darknessgate/
[+] Enjin: https://www.enjin.com/profile/darknessgate
[+] Fiverr: https://www.fiverr.com/darknessgate
[+] Freesound: https://freesound.org/people/darknessgate/
[+] G2G: https://www.g2g.com/darknessgate
[+] GaiaOnline: https://www.gaiaonline.com/profiles/darknessgate
[+] GitHub: https://www.github.com/darknessgate
[+] Gravatar: http://en.gravatar.com/darknessgate
[+] Keybase: https://keybase.io/darknessgate
[+] Kik: https://kik.me/darknessgate
[+] Linktree: https://linktr.ee/darknessgate
[+] Reddit: https://www.reddit.com/user/darknessgate
[+] ReverbNation: https://www.reverbnation.com/darknessgate
[+] Roblox: https://www.roblox.com/user.aspx?username=darknessgate
[+] Scribd: https://www.scribd.com/darknessgate
[+] SoundCloud: https://soundcloud.com/darknessgate
[+] Telegram: https://t.me/darknessgate
```

FIGURE 8.2
Using the Sherlock tool to search for a particular username on major social media platforms.

investigate the physical location of a particular mailing address, or to search for a particular geographical location found within an image or video metadata. There are different tools for aiding OSINT researchers in analyzing geospatial data to analyze physical locations. Here are the most popular ones:

• Google Maps (https://www.google.com/maps): Google Maps is a web-based mapping service developed by Google. It allows users to view maps, driving directions, and search for physical locations and businesses worldwide. Users can also access real-time traffic information, satellite imagery, and 360-degree street views of various locations worldwide. Google Maps provides a good feature for OSINT gathers: historical imagery. You can use this feature by clicking the "historical imagery" icon on the bottom left of the Google Earth desktop application. The timeline displays markings indicating the dates on which images of your map can be accessed. To examine various periods, you can adjust the timeline by dragging either the range marker to the left or right to change the time span or the time slider to shift the time range either earlier or later.

- Bing Maps (https://www.bing.com/maps): This is Microsoft mapping service. It provides different mapping visuals, such as: road, dark road, hybrid, aerial, bird eye, and StreetSide. Unlike the Aerial view which is captured by satellites, the Bird Eye view provides angled (45°) imagery from low-flying aircraft and gives high-resolution images.
- Satellites Pro (https://satellites.pro): This service allows switching between different mapping services, such as Google Map, Apple Map, Yandex Map, and OpenStreetMap.
- OpenStreetMap (https://www.openstreetmap.org): Free and open-source mapping service created by a community of mappers and aims to provide data about roads, trails, cafés, railway stations, and more, worldwide.

OSINT.link (www.OSINT.link)

OSINT.link contains an index of OSINT tools and online services grouped into categories according to their functions (see Figure 8.3).

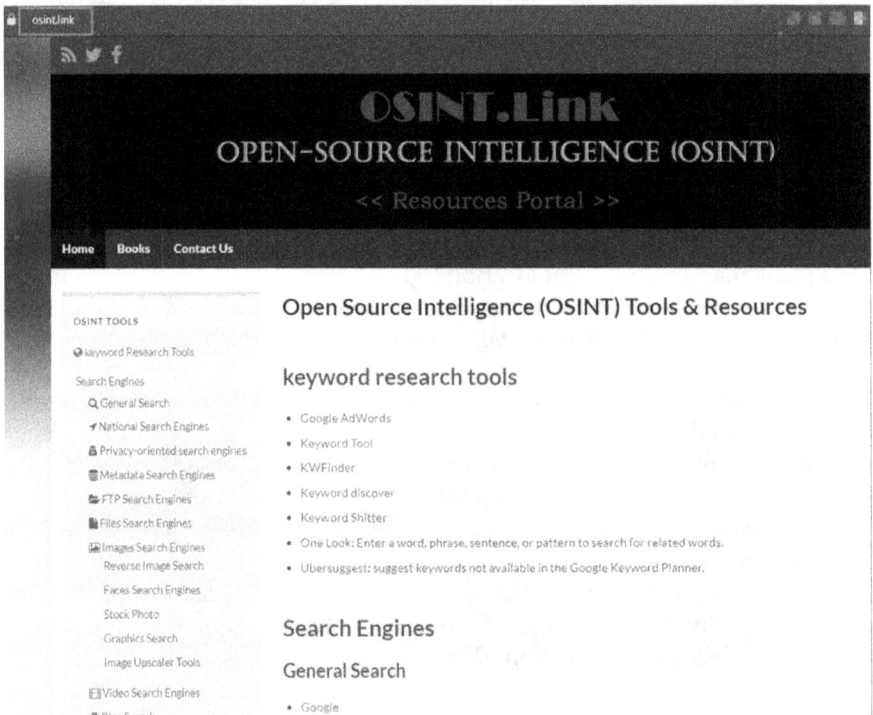

FIGURE 8.3
OSINT.link contains hundreds of Links to various OSINT tools and resources.

Open Source Intelligence Methodology

OSINT methodology refers to the process used to collect, analyze, evaluate, and disseminate information from publicly available sources. To ensure an effective OSINT gathering activity, a plan or map should be developed to guide each phase of the collection process.

The OSINT methodology entails the utilization of various tools, online services, search engines, and other techniques to collect and analyze data from diverse sources. Additionally, it requires the application of critical thinking and analytical techniques to identify and extract relevant information to build upon.

In general, an OSINT methodology is composed of the following key phases:

- **Planning and Preparation**: This is the elementary phase where the gathering objectives and the intelligence requirements are defined to guide the OSINT gatherers during the subsequent activities. During this phase, the following will be defined:
  - Determine information gathering requirements – for example, what information is required as a part of this research
  - Identify the sources that you are going to use (e.g., public databases, historical imagery maps, Wayback Machine)
  - Identify the tools needed to collect and analyze data (e.g., using *Sherlock* (https://github.com/sherlock-project/sherlock) to find connected social media accounts, using the *Harvester* (https://github.com/laramies/theHarvester) to collect domain-related information, using Gephi (https://gephi.org) to visualize data).
  - Ensure that the data that you are going to gather are legally accessible and do not violate any data protection laws
  - Determine needed resources and time constraints
- **Collection**: In the collection phase, the actual information-gathering activities begin. It includes collecting information from various sources, such as social media platforms, government databases, job websites, mapping services, historical records, Grey Literature, and other open source repositories. The goal is to collect as much relevant information about the subject entity (individual or organization) as possible. The OSINT researcher should verify the accuracy of gathered information to ensure it is reputable and can be built upon.
- **Processing and Analysis**: In this phase, the collected information is analyzed and inspected carefully using various analytical tools and techniques to discover key facts and identify patterns in data, trends, and anomalies that could help investigate the case. Collected data should be verified for the last time to ensure accuracy and no misleading information should be used to build the final report.

- **Reporting**: The final stage of the OSINT methodology involves presenting the findings to the requester in a final report or presentation. To ensure that the intelligence is effectively communicated, data visualization tools may be employed to represent conclusions in a way that is easy for decision-makers to understand. When dealing with technical data, such as cybersecurity and threat intelligence data, it is important to present the information in a language that all stakeholders, including law enforcement and senior management, can understand. The primary objective of the Production phase is to translate the insights obtained from the OSINT methodology into actionable intelligence that can support decision-making. By presenting intelligence clearly and concisely, decision-makers are better equipped to take appropriate action based on the intelligence presented.

- **Evaluation**: This phase comes after finishing delivering the final intelligence product. It is a kind of quality assurance work where OSINT gatherers discuss the OSINT project and suggest areas of improvement. For example, they discuss if the tools used were enough to collect information and if the sources used were adequate to finish the task. The evaluation phase is essential to ensure the quality of the OSINT work and to continually improve the used OSINT methodology to produce more accurate results in the future.

OSINT Challenges

OSINT offers a convenient and cost-effective way to collect intelligence from publicly accessible sources. However, leveraging OSINT also comes with several challenges; we will discuss them in this section.

Legal Consideration

Since the implementation of GDPR, organizations worldwide have become more cautious about collecting, processing, and storing individuals' personal data. Therefore, organizations must exercise great care when obtaining personal information about EU citizens from public sources as part of their OSINT collection activity. The same applies to citizens of other countries with similar data protection rules.

Limited Accessibility to Information

Some types of OSINT information require subscription and payment in order to access (e.g., grey literature materials), while some types may require specific access privileges to view the information (e.g., such as some court records information).

Enormous Data Volume

The widespread use of the internet and social media platforms has resulted in an enormous amount of digital data generated by users' daily interactions. It is now common to see internet users with multiple social media accounts. According to Demandsage,[6] the typical social media user interacts with 6.6 social media platforms.

The vast volume of online digital data requires OSINT researchers to apply different tools and techniques to filter and analyze only the most relevant information. Fortunately, the advance of AI and ML technologies can provide excellent aid for OSINT gatherers to sift through vast amounts of data (see Figure 8.4).

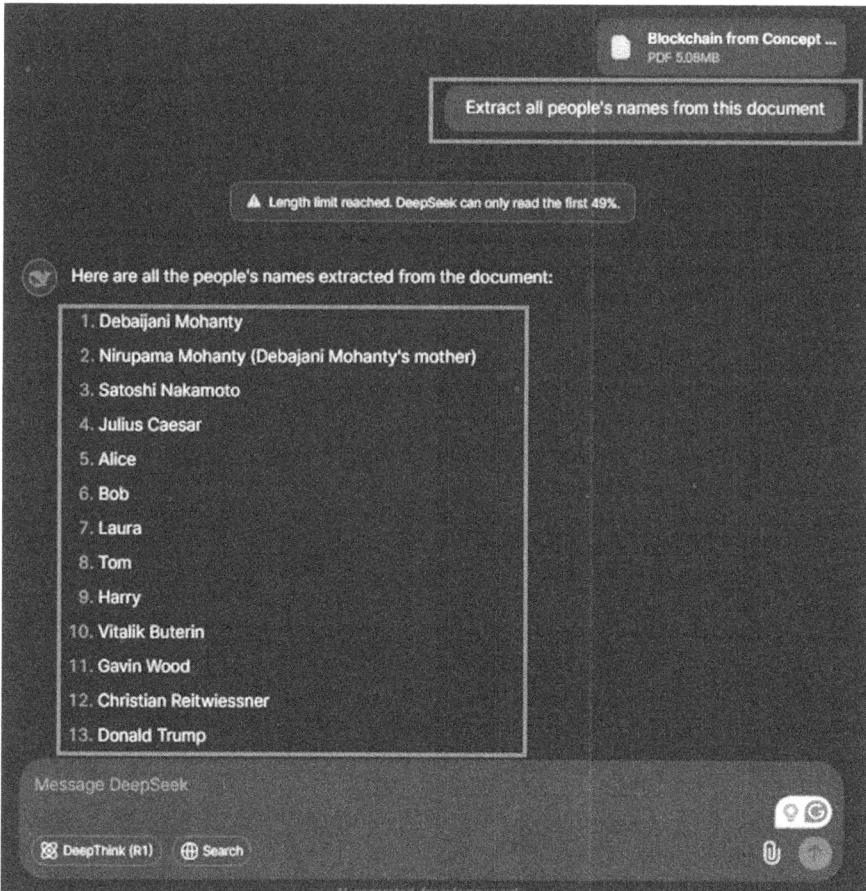

FIGURE 8.4
Using DeepSeek to extract people's names from a PDF document.

Overlapping Categories

OSINT requires researching different data sources, such as geospatial and social media information. This makes categorizing collected information into groups a challenging task. For example, geolocation information may be found in social media posts and images, which fall under both Geospatial Intelligence (GEOINT) and Social Media Intelligence (SOCMINT).

Disinformation

The proliferation of social media platforms has made it easy for people to share misinformation and fake news. According to Techjury,[7] 62% of all Internet information can be fake, and 80% of US adults have consumed fake news. For OSINT gatherers, it can be challenging to distinguish between genuine and fabricated information, especially considering the vast amount of data available online.

On the other hand, criminal and terrorist organizations could launch misinformation campaigns to mislead OSINT researchers. This requires OSINT research to verify their information sources very well before considering them accurate.

Language Barriers

Information is available online in various languages, which can make analyzing foreign language information challenging for OSINT gatherers. Translating the information is often necessary and requires extra work. Although many online translation services are available (e.g., Google Translate), understanding local idioms and expressions can be difficult for non-native speakers. Again, the advance of AI technologies will greatly aid in translating foreign languages into English promptly and help understand different idioms.

These challenges show that OSINT gathering requires more than just basic search engine skills; it involves a range of analytical and research skills. In addition, OSINT gatherers must also possess the ability to accurately translate foreign languages and understand them within their true context.

Summary

This chapter presents an in-depth introduction to OSINT gathering, which is the process of collecting publicly available information. The chapter explores the various sources of OSINT, intelligence-gathering types, and the user groups that leverage OSINT in their work. It also delves into the techniques and tools used for OSINT gathering and the challenges that come with it, such as the overwhelming volume of data and the need for source verification. The chapter details the different phases of the OSINT process, including planning and preparation, collection, processing and analysis, reporting, and evaluation. The chapter emphasizes the significance of analytical and research skills in OSINT gathering and highlights the importance of accurate translation of foreign languages. Overall, the chapter offers a comprehensive overview of the fundamental concepts and practices of OSINT gathering.

Notes

1 pentest-standard, "Intelligence Gathering", Accessed 2025-04-02. http://www.pentest-standard.org/index.php/Intelligence_Gathering
2 The International Consortium of Investigative Journalists, "Pandora Papers", Accessed 2025-04-02. https://www.icij.org/investigations/pandora-papers
3 Investorbrandnetwork, "Social Media Statistics & FAQs", Accessed 2025-04-02. https://www.investorbrandnetwork.com/solutions/social-media/social-media-statistics-faqs
4 Statista, "Number of Internet of Things (IoT) Connections Worldwide from 2022 to 2023, with Forecasts from 2024 to 2033", Accessed 2025-04-02. https://www.statista.com/statistics/1183457/iot-connected-devices-worldwide
5 Demandsage, "Social Media Users in the World – (2023 Demographics)", Accessed 2023-05-05. https://www.demandsage.com/social-media-users
6 Demandsage, "Social Media Users in the World – (2023 Demographics)", Accessed 2025-04-01. https://www.demandsage.com/social-media-users
7 Techjury, "18 Eye-Opening Fake News Statistics for 2023", Accessed 2025-04-01. https://techjury.net/blog/fake-news-statistics/#gref

9
Endpoint Defense Strategies

Introduction

In today's world, computer networks are fortified with various perimeter-based security measures to guard against cyberattacks. However, cyber-criminals are constantly seeking weak entry points to infiltrate targeted networks, and one such point is often found in the security vulnerabilities of end-users' endpoint devices.

For large enterprise networks, a breach of even a single endpoint device can result in catastrophic consequences for the organization. Cybercriminals can use a compromised endpoint as a gateway to the entire network, allowing them to move laterally, plant malware, steal sensitive data, and carry out other malicious activities.

An endpoint device is any computing device that can connect to a computer network to provide access to network resources. Under this definition, the following are common types of endpoint devices:

- Computer workstations and servers
- Laptops
- Point-of-sale (POS) systems
- Network-attached storage (NAS) devices
- Virtua machines
- Mobile devices – such as tablets, smartphones
- Printers and scanners
- Phones with Voice over IP (VoIP) technology
- Security cameras
- Smart TV and intelligent home devices such as smart refrigerators, locks, lights, and thermostats. These devices fall under the Internet of Things (IoT) devices category

Endpoint devices serve as gateways to an organization's IT environment. Threat actors are aware of this fact and actively seek to exploit these

DOI: 10.1201/9781003008279-9

devices in order to gain access to the target environment. Recent research by Adaptiva[1] found that the average enterprise now manages approximately 135,000 endpoint devices. This is a big number and is expected to increase rapidly as digital transformation continues its boost worldwide to include more IoT within organizations' IT environments.

This chapter is dedicated to discussing the various risks associated with endpoint devices and the best prevention methods to counter them. However, before we begin, let us define the term "Endpoint Device Security".

What Is Endpoint Security?

Endpoint device security refers to the process, technologies, and practices used to secure endpoint devices from malware, ransomware, phishing, and other cyber threats. Endpoint device security is part of any organization's overall cybersecurity strategy, as compromising a single device can jeopardize the entire network's security. Securing endpoint devices is vital for safeguarding your organization's data.

Endpoint device security has become more critical recently due to the following reasons:

- The workforce has become more distributed since the COVID-19 pandemic. Today, organizations operate using one of three schemes: office-based, remote, and hybrid. Remote and hybrid workers often rely on their personal devices to access company resources. End-user devices are typically less secure than devices used in office settings because those users may not have the technical expertise to configure their devices for security.

- Data breaches become very costly. The global average cost of a data breach has reached $4.35 million, while it reached $9.44 million in the United States in 2022.[2]

- The average cost of endpoint attacks has reached $8.94 million.[3] This significant expense is primarily attributed to the loss of IT and end-user productivity, as well as the theft of information assets.

- The number of IoT devices is increasing rapidly due to the adoption of 5G (and later 6G, which is 100 times faster than 6G) and other advances in communications technologies. The number of installed IoT devices worldwide is expected to reach 25.44 billion by 2030.[4]

- In a recent survey, 600 CISOS, IT SecOps Directors, and Managers have expressed their worries about endpoint device security – 50% of endpoints are at risk.[5]

Gartner predicts that in 2023, 39% of global knowledge workers will be working in hybrid fields. Endpoint device security is critical as endpoint security breaches can encompass a range of outcomes, such as financial loss, compromise of customer personally identifiable information (PII), reduced productivity of IT and end-users, system outages, theft of valuable information assets, and damage to business reputation.

Another study by Upwork finds that 22% of the American workforce will become remote by 2025.[6]

Endpoint Security Risks

Exposing an endpoint device to security risks will not only affect the device. For instance, it could compromise the entire IT environment of the impacted organization. The following is a list of the most common attacks against endpoint devices.

Malware

Malware can be introduced to endpoint devices via various attack vectors, such as phishing emails, malicious attachments, malicious USB devices, and by visiting compromised websites housing malicious code (e.g., exploit kits). As we saw in Chapter 4, there are various types of malware according to their malicious behavior, such as ransomware, keyloggers, backdoors, and malvertising. Malware can be used to steal data, remotely control the infected device, and use it later to move laterally across the targeted organization's network to conduct various malicious actions. Malware is among the top attack vectors utilized by cybercriminals to attack endpoint devices.[7]

Phishing Attacks

Hackers employ social engineering (SE) techniques to deceive victims via email, text messages, or in-person interactions. They aim to trick targets into revealing sensitive information such as login credentials and credit card information. Through SE tactics, victims may also be tricked into downloading malware onto their endpoint devices. The credentials and access harvested via SE provide hackers with initial access to compromise endpoint machines.

Once a foothold is established on the target end-user device, threat actors can pivot to further infiltrate the broader organizational network. The infected endpoint provides a gateway to move laterally and escalate the attack. Additional malicious actions can then be conducted against the organization's IT environment, where the hijacked endpoint system is connected.

DDoS Attacks

Endpoint devices can be affected by a DDoS attack in two ways. First, the endpoint device itself can be targeted, rendering it unavailable to legitimate users. Second, a compromised endpoint device can be enlisted in a large botnet network to launch DDoS attacks against other targets.

Insider Threats

Insider threats are any threat originating from people within the organization. Not all insider threats are malicious. For example, neglected employees may infect their endpoint devices with malware after installing malicious internet programs or accessing pirated software websites (such as torrent websites). On the other hand, malicious insiders, such as disgruntled employees and third-party contractors with legitimate access to endpoint devices and network resources, can conduct various malicious actions, such as data theft and sabotage.

Software Vulnerabilities

Software vulnerabilities are generally defined as any security weakness in the software running on endpoint devices. It can be caused by different issues, such as:

- Wrong implementation of software design
- Wrong configuration in the software program
- Wrong configuration in the operating system (OS) where the application is running
- Security flaws in the software dependencies of the application – for example, the connected database system

These vulnerabilities pose a direct security threat to endpoint devices. Cybercriminals can exploit software weaknesses in different ways:

- Gain unauthorized access to the endpoint device to steal data (e.g., personal data, login credentials, or financial details) or hijack the system.
- Plant malware, such as ransomware, in the infected endpoint device and use the infected device as a basis to spread the infection to other devices across the network.
- Software vulnerabilities can be used to install specific malware to facilitate executing Denial-of-Service (DoS) attacks.

It is worth noting that software vulnerabilities are not merely limited to those that exist on applications running on endpoint devices. For instance, security vulnerabilities can be broadly grouped into four main groups:

- Network vulnerabilities
- OS vulnerabilities
- Process vulnerabilities
- Human vulnerabilities

Weak Passwords

Using weak passwords that are easy to guess and predicate by employees will expose their endpoint devices and network accounts to risk. For example, weak passwords can be cracked relatively easily using a dedicated password cracker tool, such as Cain & Abel[8] and John the Ripper[9]

Bring Your Own Device

Bring Your Own Device (BYOD) is a practice that allows employees to use their personal devices for business purposes. This practice has increased since the COVID-19 pandemic and the increased hybrid work scheme. Using personal devices at work can increase employee satisfaction and productivity. However, it also introduces significant security risks such as:

- Theft of devices containing sensitive work data.
- Employee may install malicious applications by mistake or visit compromised websites housing exploit kits that can introduce malware into their devices.
- Limited visibility and control of the organization's IT department. Personal devices cannot be secured to the same extent (e.g., installing security solutions and enforcing strict access controls) and monitored at the same level as company-owned devices.

Lack of Encryption

Encryption safeguards data from unauthorized access, both when it's stored (at rest) and when it's being transmitted (in transit). However, if the hard disk of an endpoint device is not encrypted, it becomes vulnerable to threat actors who gain physical or remote access to the device. Similarly, if data packets transmitted between the endpoint device and other devices on the network are not properly encrypted, they can be intercepted by threat actors monitoring network traffic.

Portable endpoint devices like laptops and tablets are particularly suscep-tible to theft and loss. Without encryption, all the data on a lost device can be accessed by unauthorized parties.

Physical Risks

We have already indirectly covered the physical risks of endpoint devices. For instance, portable endpoint devices are subject to theft and loss. When employees use their personal devices to access corporate data, leaving them unattended in public places creates an opportunity for threat actors to gain unauthorized access to these devices.

Endpoint Security Technologies and Best Practices

Now, it's time to talk about the essential measures, tools, and strategies to establish robust endpoint device security against cyber threats. These comprehensive strategies aim to safeguard both the data at rest on end-point devices and the data transmitted during interactions with other devices. By implementing these practices, organizations can effectively mitigate the risks posed by cyber threats and enhance overall endpoint security.

Install Antivirus and Antimalware

Antivirus and antimalware solutions are crucial for protecting endpoint devices from computer viruses and other malware. Antivirus software detects and stops malicious code found in websites and files, providing defense against traditional threats like worms, viruses, trojans, spyware, and keyloggers before they infect your computer. However, it is essential to note that antivirus alone cannot offer complete protection against advanced malware, including ransomware, rootkits, and other sophisticated threats. That's why installing an antimalware solution on your endpoint device is also essential.

An antimalware program complements antivirus software by capturing threats that may have been missed, such as ransomware, and preventing their rapid spread to other devices on the network. Some antimalware pro-grams utilize proactive measures, like behavioral analysis, to detect malware based on its malicious behavior and actions within the system rather than relying solely on signatures. This enables the immediate blocking of ransom-ware before it can initiate its encryption routine, providing enhanced protec-tion for data stored on endpoint devices.

Install Personal Firewall

Firewalls are either hardware devices or software programs used to regulate access to network resources. For endpoint devices, it is advisable to employ personal firewall software programs that control data traffic flow into or out of the device. A personal firewall helps prevent intrusion attempts from the internet or the local network.

Installing a dedicated personal firewall on endpoint devices is vital for the following reasons:

- Personal firewalls offer an extra layer of security beyond network perimeter firewalls. They monitor incoming and outgoing traffic on endpoints, instantly identifying and blocking suspicious packets.
- Prevents intrusions by blocking attempts to exploit security vulnerabilities in endpoint OSs or in the installed applications.
- Endpoint device firewalls allow managing which applications can run or access specific network resources or the internet.
- Personal firewalls prevent intrusion attempts when connecting your endpoint device to a public WiFi hotspot. Attackers on the same network can try to access your device, and a firewall will help halt such attempts.
- Personal firewalls help prevent malware infections by blocking access to compromised websites and preventing suspicious network activities from accessing your endpoint device resources.
- Installing a firewall on endpoint devices has become a requirement for some compliance regulations, such as Payment Card Industry Data Security Standard (PCI DSS) – Requirement 1.1.6[10]

Note: You cannot install a firewall on all endpoint device types. For instance, some IoT devices, such as wearable devices and sensors, do not have the resources (e.g., memory and processing power) and technical capabilities to run a dedicated personal firewall.

Install Endpoint Detection and Response

Endpoint detection and response (EDR) is an intelligent endpoint security solution used to detect advanced threats using advanced detection mechanisms, like behavioral analytics and machine learning (ML) technologies. EDR records users' normal behaviors while using their endpoint devices (e.g., users' login, process execution, communications with internet websites, and other endpoint devices across the network) and compares them to current user behaviors. This allows for the detection of abnormal activities and the automatic blocking of them.

Unlike other security solutions, EDR can detect advanced threats that surpass traditional network perimeter defenses and respond to them immediately. Some EDR solutions come connected with a threat intelligence feed to detect zero-day vulnerabilities and emerging threats, which can effectively help maintain the security of endpoint devices from zero-day attacks.

Different EDR vendors are in the market. Some sell their EDR solution to customers to manage it, while others provide EDR-managed service. It is always advisable to select the EDR solution that provides the best technical capabilities, such as:

- Utilize advanced detection capabilities: Select an EDR with capabilities powered by machine behavioral analytics, ML, and artificial intelligence (AI) algorithms to automate threat identification.

- Can integrate with threat intelligence feeds: Threat intelligence feeds foster the EDR capability to detect known threats by providing current threat information such as malicious IPs, domain names, and other indicators of compromise (IOC).

- Forensic capability: An EDR solution with digital forensics capabilities assists in creating a timeline of activities on compromised endpoint devices both before and after a security breach. This capability effectively supports forensic investigators in identifying the root cause of security incidents.

- Full visibility: The EDR solution should provide comprehensive visibility of all endpoint device interactions, such as processes running, network connections, file activities, and any changes in system configurations.

- System resources: Select the EDR that consumes minimal system and network resources so it will not degrade endpoint devices' performance.

- Integration with other solutions: EDR solution should be able to integrate with existing security solutions, such as SIEM (Security Information and Event Management) systems and security orchestration platforms.

- Reporting and auditing: If your organization must adhere to specific compliance requirements, such as PCI DSS, GDPR, or HIPAA, ensure that your EDR solution includes the necessary features, such as reporting and auditing capabilities, to facilitate meeting these regulatory obligations.

Install Anti-Phishing Solution

Anti-phishing technology is developed to prevent phishing attacks from entering an organization's email system. Phishing attacks trick users into sharing sensitive information or clicking on harmful links. Phishing is considered among the most dangerous cyberattacks against individuals and organizations. According to IBM,[11] Phishing accounted for 16% of cybercrime attack vectors, leading to an average breach cost of $4.91 million.

Anti-phishing solutions can be either network-based, which resides on the network perimeter or host-based, installed on the endpoint device.

Some antivirus solutions have an integrated anti-phishing capability, while some vendors produce standalone programs for endpoint devices. A modern approach to handling anti-phishing is using cloud-based solutions managed by a third-party provider. Some cloud-based solutions integrate seemingly with Office 365 and G Suite.

Anti-phishing solutions work by scanning incoming emails for:

- Signs of phishing are identified by analyzing the language used in the email.
- Scan links contained within incoming emails and attachments to identify suspicious websites and attachments and prevent end users from accessing them.

Some advanced anti-phishing solutions come with enhanced security features, such as:

- Connect with threat intelligence feeds to stay current on the latest phishing attacks, trends, and techniques, allowing them to recognize emerging phishing attacks instantly.
- Utilize ML technology to analyze email patterns to detect phishing emails before they access end-user inboxes.

There is no single technological solution that can prevent phishing attacks 100%. The best method to mitigate phishing is to adopt a hybrid approach combining the latest anti-phishing technology on the network perimeter and endpoint devices in addition to taking care of end-user cybersecurity awareness training.

Note! Popular Threat Intelligence Feeds

- AlienVault (https://otx.alienvault.com)
- The InfraGard Portal by the FBI (https://www.infragard.org)
- SANS: Internet Storm Center (https://isc.sans.edu)
- VirusShare (https://virusshare.com)

Threat Extraction

Threat actors are increasingly using malicious documents delivered via emails to infect with malware. For instance:

- Over 94% of malware is delivered via email[12]
- 48% of malicious email attachments are Microsoft Office Files[13]
- PDF files count for 14% of total malicious file extensions[14]

Historically, many devastating malware attacks leveraged malicious documents as the attack vector; the following are the most notable cyberattacks that leverage malicious documents to deliver the malware:

- Emotel: A banking trojan delivered via malicious MS Office files, specifically MS Word and MS Excel
- Jaff ransomware: Appeared in 2017, it leverages malicious PDF files
- Locky ransomware: Use malicious MS Word files sent via email
- APT MuddyWater malware: Use malicious macros embedded within MS Word files to infect target devices with malware

Threat extraction (also known as file sanitization or CDR (content disarm and reconstruction)) is a security technology that removes exploitable content from documents, such as PDF and MS Office files.

The CDR technology does not rely on detection techniques like virus signatures to detect threats. Instead, it assumes all files are malicious and should be scrutinized first before opening them. The CRD works by removing embedded malicious code in PDF documents, OpenDocument text format (.odt), malicious Office Macros, or other file types, and finally rebuilding the document again to maintain its integrity.

Attackers leverage file attacks to infiltrate organizations' defenses by attaching malicious documents such as PDF files and MS Office documents and sending them via email or Internet Messaging (IM) applications such as WhatsApp or Slack. When an unaware user opens the file, it executes the malicious code to deliver the payload.

CDR (Content Disarm and Reconstruction) solutions are typically deployed at network perimeters, such as email gateways and web proxies. However, certain CDR solutions can also be installed on endpoint devices to provide additional security. Implementing a CDR solution on the endpoint makes sanitizing malicious code within PDFs, MS Office, and archive files possible before the user opens them. This proactive approach significantly reduces the various risks associated with file-based attacks, bolstering endpoint security.

Vulnerability Management

Vulnerability management is a proactive process involving identifying, prioritizing, and mitigating vulnerabilities in endpoint devices and other critical systems (such as virtual machines and networking devices) across an organization's IT environment. The primary objective is to prevent threat actors from exploiting these vulnerabilities for malicious purposes, ensuring the overall security of the organization's digital assets.

Vulnerability management is not something your organization should do once or twice per year. It is a continuous process of discovering new vulnerabilities, prioritizing them according to their severity, and acting accordingly to fix them before they get exploited by hackers. Every organization should own a vulnerability management program, which must be integral to its cybersecurity strategy.

> Note! Although vulnerability management is covered in the next chapter on enterprise network defenses, patch management is also mentioned here due to its critical importance for endpoint security.
>
> Patch management falls under the broader discipline of vulnerability management. However, regularly patching endpoint devices to fix software flaws is a fundamental security practice that deserves dedicated discussion in this endpoint security chapter.

Vulnerability management is not a single process; it is composed of the following four main processes:

1. Vulnerability scanning – Your security team should scan organization IT assets – including cloud services – once every week to ensure emerging vulnerabilities are not left unhandled. Nessus is a popular vulnerability-scanning tool.
2. Vulnerability assessment – This should be done every week. This process aims to assign severity levels for identified vulnerabilities. Common tools used in this process are Nmap and Burpsuite.
3. Patch management – This should be done every day if applicable. This process aims to distribute updates to OSs and third-party software components. More about Patch Management will be discussed in the next section.
4. Vulnerability remediation – This should be done every week. This process aims to remediate discovered vulnerabilities in software applications and networks and ensure the remediation is successful and the applied fix will last for a long time.

Patch Management

Patch management is the process of applying software updates to various endpoints and other networking equipment, such as routers, switches, and Firewalls. For instance, OSs, applications, and software running on networking devices should be kept current and patched regularly to close open security vulnerabilities and ensure all devices within your digital ecosystem are not exploitable by threat actors.

For endpoint devices, patch management is critical for the following reasons:

- Fix or close security vulnerabilities in software components within your IT environment. Patch management ensures your endpoint devices receive the latest security patches and updates from all your software vendors.
- Increase system uptime and prevent sudden outages. Exposing your endpoint to cyber threats may make them unavailable, which results in losing employee productivity.
- Reduce the risk of malware infections. Many malware attacks exploit outdated or unpatched OSs or applications to infiltrate endpoint systems.
- Enhance endpoint device performance, as not all updates are security-related. For instance, updating your software products to the latest version will ensure you receive the latest functionality and enhanced performance of the modern versions.
- Patch management has become required for many regulatory compliance bodies, such as PCI DSS Requirements 6.1, 6.2, and 6.3.

There are different patch management solutions to automate the process of patching OSs and applications for endpoint devices. Here is a list of the most popular ones:

- Microsoft Windows Server Update Services (WSUS)
- SolarWinds® Patch Manager
- Ivanti Patch Management
- IBM BigFix Patch Management

When choosing a patch management solution, it is important to consider the following criteria:

- Compatibility: Ensure the solution is compatible with your IT ecosystem, including the OSs and applications used in your environment.

- User interface: Select a solution that offers an intuitive and easy-to-use user interface.

- Scalability: Consider whether the solution can scale according to the number of endpoint devices in your IT environment. It should handle the growth or reduction in device count effectively.

- Reporting capabilities: Look for a solution that provides comprehensive reporting capabilities. This will help you to generate compliance reports.

- Third-party software patching: Verify if the solution supports patching for third-party software applications. This ensures that the OS updates are covered and that critical patches are provided for commonly used applications.

- Support different OS: The patch management solution should support major OSs such as Windows, Mac, and Linux. This ensures all endpoint devices across your IT environment are receiving proper updates regardless of their OS type.

- Integration: Check if the solution can integrate with other security solutions, such as Security Information and Event Management (SIEM) systems. The integration allows for a cohesive security strategy and centralized management.

Encrypt Endpoint Devices' Data

Endpoint devices often store sensitive business data. Though they are not standalone entities, they need to interact with other devices and cloud services in your IT environment for data exchange. Moreover, portable devices like smartphones, laptops, and tablets are prone to physical security risks such as loss or theft. To safeguard the security and integrity of endpoint device files, it is crucial to employ robust encryption for their protection.

Data that must be secured at endpoint devices include data at rest (stored data) and at transit (transmitted data – for example, when interacting with cloud services or accessing corporate resources remotely).

Numerous encryption algorithms are on the market today; the most popular – and free – ones are Advanced Encryption Standard (AES) and Rivest-Shamir-Adleman (RSA).

Encrypting data at endpoint devices is very important for the following reasons:

- Data protection: Endpoint devices often store or process sensitive customer data, including personally identifiable information (PII), patients' health information, and financial data. Unauthorized access to this information can lead to severe consequences for the affected organization. Encryption helps safeguard this data, making it unreadable to unauthorized parties.

- Insider threat prevention: Endpoint devices are susceptible to insider threats, where disgruntled employees or third-party contractors with legitimate access may attempt to access sensitive information. Encryption keeps the data protected and inaccessible to threats, even if unauthorized physical access occurs.

- Compliance requirements: Many compliance regulations, such as GDPR, PCI DSS, and HIPAA, require organizations to implement specific measures to protect sensitive customer data. Encryption is often a key requirement to meet these compliance standards and ensure data security.

- Mitigating data breaches: Encrypting stored information and data during transit is a preventive measure against data breaches. It significantly reduces the risk of threat actors intercepting or accessing the data, thereby enhancing overall data security.

- Enhance organization security exposure: Employing encryption will increase customer trust in your services. It will also increase third-party contractors' and business partners' confidence in the security of your IT systems as they will guarantee that their sensitive information is saved and processed using top security measures.

Endpoint device encryption can be applied in two ways.

Full-disk Encryption

In full disk encryption (FDE) (also known as whole disk encryption), all data stored on the endpoint device's hard drive is encrypted, including swap files, system files, temporary files, and hibernation files. Remember that the FDE will not protect your data if you log into your computer, decrypt the data, and leave your computer unattended. Unauthorized users can access your data while in a decrypted state.

FDE is required when you want to protect all files and applications stored on an endpoint device. This is advisable for laptops containing sensitive work documents or used to access corporate sensitive applications.

If your endpoint runs Windows, you can use BitLocker's built-in full-volume encryption feature. It is worth noting that BitLocker is not available in all Windows versions. For instance, it is supported on the following Windows versions:

- Ultimate and Enterprise editions of Windows Vista and Windows 7
- Pro and Enterprise editions of Windows 8 and 8.1
- Pro, Enterprise, and Education editions of Windows 10
- Pro, Enterprise, and Education editions of Windows 11
- Windows Server 2008 and later

A popular open-source program for conducting FDE is *Veracrypt* (https://www.veracrypt.fr/code/VeraCrypt), which supports major OSs.

There is also another open-source program for encrypting your files on the client side before sending them to the cloud. It is called Cryptomator (https://cryptomator.org), and it can run on all major mobile OSs, such as Android and iOS. We can also install it on major desktop OSs such as Windows, Linux, and macOS.

> **Note!** Implementing FDE on portable hard drives, such as USBs and external hard drives, should be mandatory.

File Encryption

Unlike FDE, which encrypts all content available on the hard drive, including the installed OS, files or folders are encrypted individually in file encryption. This requires user action. For example, selecting which file/folder to encrypt or decrypt and entering the password or the secret key file to decrypt the data.

While FDE can be implemented on the hardware or software level and requires more computing power, file encryption is executed at the software level and can be performed relatively easily by any computer user.

Many software security vendors provide built-in encryption functions as a part of their general endpoint security solution. Still, there are many reputable free programs for conducting file/folder encryption, such as:

- Gpg4win (https://www.gpg4win.org), which can be used for file and email encryption.
- 7-Zip (https://www.7-zip.org), which is a file archiving program; however, we can use it to encrypt individual files/folders as well (see Figure 9.1).

Encrypt Data in Transit

Endpoint devices are not working in isolation; they need to interact with cloud services or with other devices across the organization's IT environment. Failing to encrypt moving data will allow threat actors to intercept them. Here are some methods to safeguard moving data across computer networks:

- Transport Layer Security (TLS)/Secure Sockets Layer (SSL): These protocols guarantee that your transmitted data from endpoint devices will get scrambled. Make sure your cloud provider supports securing connections via TLS/SSL.

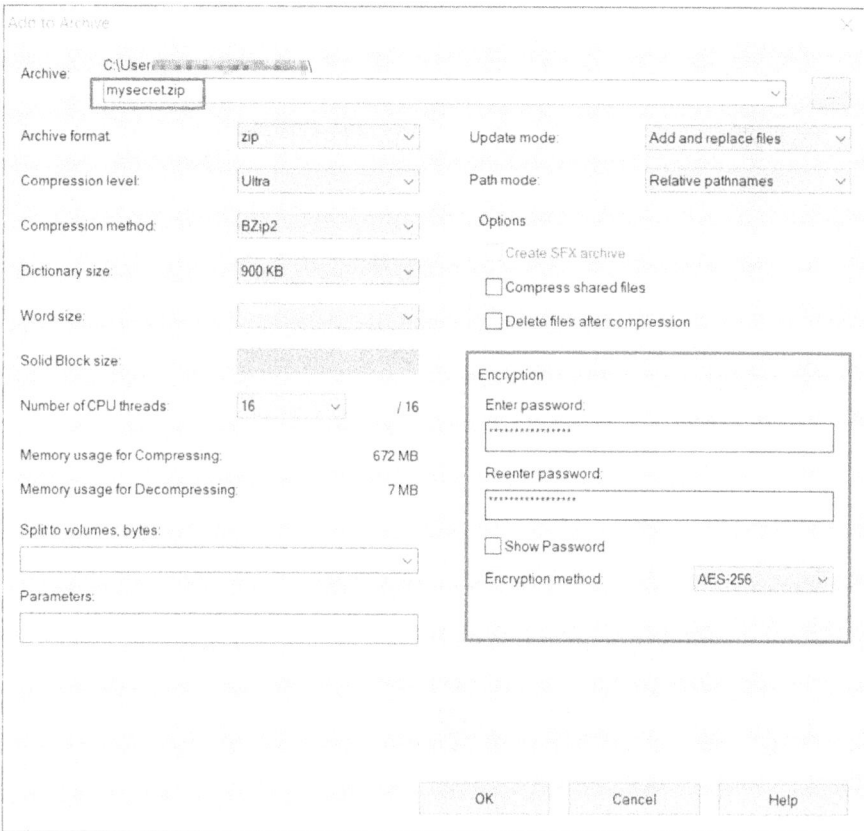

FIGURE 9.1
7-Zip can be used to encrypt individual files/folders.

- Virtual Private Network (VPN): A VPN technology creates a secure virtual tunnel between two points or devices across computer networks. It is widely used to secure endpoint connections to organizations' remote protected resources or cloud services across the internet. All data transmitted through the VPN is encrypted and protected from third-party interception. VPN helps achieve data privacy, as it can conceal your device IP address from third-party observers, such as your internet service provider (ISP) and advertisers.

Note! Cloud providers typically offer encryption capabilities to their clients. Refer to the documentation provided by your cloud provider to learn how to implement encryption when transferring data between your cloud infrastructure and endpoint devices.

Enable Multi-factor Authentication

In Chapter 2, we discussed encryption in detail and discovered that relying solely on traditional passwords is no longer enough to secure access to your endpoint devices, cloud applications, and other online accounts. Two-factor authentication (2FA) requires a user to use more than one authentication factor to access protected resources. For example, a user's traditional password and a temporary passcode are sent via text message (SMS) or email.

2FA protects against different types of password attacks, such as brute force, password spraying, credential stuffing, and dictionary attacks. Weak passwords are still the main vehicle used by hackers to infiltrate IT systems. For instance, the Verizon 2022 Data Breach Investigations Report[15] found that weak or stolen credentials account for a significant majority, specifically 81%, of data breaches caused by hacking activities.

Use Identity and Access Management Solution

The primary security concern for organizations today revolves around effectively managing user identities as they access their IT environment. Today's digital IT infrastructure spans local and cloud premises. It contains a large number of applications and IoT devices. Keeping tabs on all the identities that interact with these complex networks has become increasingly challenging.

Identity and Access Management (IAM) is a framework that defines processes, security policies, and tools used to manage users' access to protected digital resources. The main technologies used in IAM solutions are:

- Authentication methods – such as 2FA, MFA, or single sign-on (SSO)
- Privileged access management (PAM), which contains more tools such as a secure vault to keep users' access credentials and session tracking to track users' access once they have been granted access
- Security-policy enforcement applications
- Reporting applications

IAM solutions provide organizations with a centralized directory to store all users' identities and access privileges. It's important to note that managing access to protected resources extends beyond human users. Software applications, hardware devices (such as IoT devices), and other systems also require the ability to interact with the IT environment and should have access credentials that define their access permissions within the environment, similar to human users.

IAM helps mitigate many cyberattacks against endpoint devices. For example, through the implementation of MFA, IAM ensures that attackers cannot gain access to a particular endpoint device by simply knowing the user's password. IAM solution can also enforce adaptive access controls based on factors like user role, device state, IP location, and time of day. This will efficiently limit the exposure of endpoint devices to external attacks.

On the other hand, IAM safeguards applications by specifying the access roles of individual users and specifying their permissions for each application or file within the system. Many IAM solutions allow integration with threat intelligence feeds, further enhancing their ability to detect emerging threats and zero-day attacks.

Data protection regulations such as GDPR, HIPAA, and PCI DSS require organizations to monitor how users access sensitive customers information. IAM solutions can help meet this requirement by providing centralized control and visibility of user access privileges. By implementing an IAM solution, organizations can establish access controls, manage user identities, and keep track of user activities across the IT environment. This helps ensure compliance and enables organizations to monitor and generate reports on user actions, demonstrating their adherence to these regulations.

IAM solutions can be deployed locally (on-premises) or in the cloud. Many companies, especially small ones, outsource their IAM to a third-party IAM provider. Outsourced IAM is normally run in the cloud; this greatly helps reduce maintenance costs and increases uptime for their customers.

Apply the Principle of Least Privilege

The principle of least privilege (PoLP) is a security concept that entitles an entity (user, system, application, application function, or device) to only be given access to the required resources to execute their task and nothing else. By implementing the PoLP, organizations can significantly reduce their cyber-attack surfaces and prevent malware infections from spreading from one endpoint device to the entire network.

The PoLP is fundamental for protecting endpoint devices for the following reasons:

- Lower the impact of compromised accounts: Suppose attackers gain access to specific endpoint device user credentials. Implementing PoLP will – by default – give limited permissions to each user, which will significantly restrict the attackers' ability to navigate the compromised system. This mitigates the risks associated with installing malware or accessing sensitive customer information.

- Prevents privilege escalation and lateral movement: If attackers successfully gain unauthorized access to an endpoint device, by implementing the PoLP, they will not be able to move laterally across the network to infect more devices because they will have a very restrictive account, which prevents them from escalating their privilege or from compromising other systems across the network.

- Minimize cyberattack surface: Endpoint devices are lucrative targets for hackers because they are relatively easier to attack than network servers. By limiting administrative privilege (except for IT administrators who need such access to accomplish their work) from endpoint devices, attackers will become unable to execute admin actions (execute programs or wide access to other devices and shares across the network) on compromised endpoint devices, which subsequently limit their damage ability.

- Helps achieve regulatory requirements: Different regulatory compliance bodies and industry standards encourage implementing the PoLP to secure sensitive customers' information. For instance, the PCI DSS requirement 7.2 and HIPAA access control standard (45 CFR § 164.312) mandate implementing PoLP on privileged accounts to prevent unintentional or malicious damage to sensitive data.

The PoLP is a critical element of the zero-trust security strategy, which requires organizations not to automatically trust any entity trying to access their applications and data, whether the request is coming from outside or inside organization network perimeters. Every entity should be evaluated before granting access to the protected resources. The zero-trust model can be defined in two phrases: never trust and always verify.

Harden Endpoint Devices Operating Systems

System hardening refers to implementing a set of tools, configurations, and best practices to enhance its security by closing loopholes in OSs, applications, firmware, and other devices, such as IoT and networking devices. By doing so, the attack surface will be reduced significantly, making infiltrating the hardened system very difficult for threat actors.

There are different types of system hardening, such as:

- Server hardening
- Endpoint (e.g., workstation, laptops, IoT) hardening
- Network hardening
- Application hardening
- Database hardening

However, regardless of the computing system you will harden, you should follow a systematic approach that begins by identifying the security issues that threat actors could exploit (usually conducted by using a vulnerability scanner program) and applying the necessary measures to close them.

For endpoint devices, hardening involves hardening the OS and installed applications. However, as endpoint devices need to interact with other devices across the network, server and network hardening should be covered as well. The main hardening tasks include the following:

- Change default passwords (factory set credentials) of devices – such IoT devices and default accounts on servers, workstations, and laptops. Attackers use automated programs, such as IoTSeeker (https://github.com/rapid7/IoTSeeker), to find accounts that still utilize default passwords to hijack and gain a foothold in the target organization's digital ecosystem. Datarecovery has a repository of default passwords for various devices and applications (https://datarecovery.com/rd/default-passwords). Another website that lists default passwords is: https://default-password.info

- Ensure the computer OS and all installed applications are current with the latest security patches and updates. Firmware should be patched when an update is available to mitigate its security vulnerabilities, especially for IoT devices. According to Microsoft's March 2021 Security Signals,[16] the occurrence of firmware attacks is increasing, as stated in the report, with over 80% of enterprises having encountered at least one such attack within the previous two years. However, only 29% of security budgets are dedicated to safeguarding firmware.

- Uninstall applications that are not used or needed at work.

- Uninstall unnecessary drivers.

- Disable unneeded users accounts, especially those with admin access rights.

- Close unnecessary ports and stop unnecessary services and network protocols.

- Use strong and complex passwords to protect users accounts. A strong password should be at least 14 characters and contain a mixture of uppercase letters, lowercase letters, numbers, and special characters (e.g., @, $, !, %). Make sure not to use the same password to secure more than one account

- Configure your system/application to lock out users' accounts after a specific number of failed login attempts.

- Disable USB ports and prevent attaching removable storage devices with endpoint devices and servers.

- Network hardening involves:
  - Checking firewall rules and updating them regularly
  - Close unused network protocols
  - Disable legacy protocols such as Echo and Chargen
  - Implement strict access lists for end-users

- Install appropriate network security controls
- Enable SSHv3 or TLS to ensure all communications across the network is probably encrypted
- Disable unused switch ports
- Restrict access to networking devices' console port
- Restrict physical access to networking devices
- Avoid sharing network configuration files using insecure means
- Use penetration testing to test the security of your network periodically
- Application hardening involves:
  - Implementing best coding practices when developing software applications – check OWASP Secure Coding Practices Quick Reference Guide https://owasp.org/www-pdf-archive/OWASP_SCP_Quick_Reference_Guide_v2.pdf
  - Generate a software bill of materials (SBOM) for your software apps. SBOM includes a list of all software components used in building software. The USA National Telecommunications and Information Administration has published a guide for the "Minimum Elements For a Software Bill of Materials (SBOM)" https://www.ntia.doc.gov/files/ntia/publications/sbom_minimum_elements_report.pdf
  - Ensure the installed applications are configured properly for security
  - Sanitizing user input to prevent web attacks such as SQL injection, cross-site scripting (XSS), remote file inclusion (RFI), and directory traversal
  - Govern access to applications based on users' job roles – for example, by using an IAM solution
  - Force all users to utilize MFA to access applications and OS

Finally, all users' actions when accessing endpoint devices or other network resources should be logged and audited. This helps detect security incidents and simplifies the investigation of cyberattacks for digital forensics experts.

The National Institute of Standards and Technology (NIST) has a special publication dedicated to server hardening; you can find it at: https://nvlpubs.nist.gov/nistpubs/Legacy/SP/nistspecialpublication800-123.pdf

Create and Enforce Security Policies

Security policies are written documents that outline the steps an organization will undertake to protect its IT assets and data. The security policy will typically mention the different access controls, processes, and measures that should be utilized to protect IT assets and data from unauthorized access or damage.

While the security policy is not concerned with low-level technical implementation, it will provide a guide for implementing various technical security controls – such as a policy may state that 2FA should be used to access some resources or prevent employees from installing unapproved applications on their work devices.

Security Policies Types

There are different types of security policies. They differ according to each organization's work scope and requirements. The NIST[17] differentiates between three types of security policies:

Program Policy

Program policy establishes an organization's information security program, which identifies the direction of its security strategy along with the required resources to achieve it. Program policy will not change frequently because it defines the high-level strategy without mentioning technical details on how they are going to be implemented.

Issue-Specific Policy

Issue-specific policies provide guidance on security matters relevant to an organization's workflow and processes. These policies focus on providing guidelines for employees when using the organization's IT systems or accessing its data. For instance, if employees are allowed to use personal devices at work, a BYOD policy should be in place. Similarly, a security policy may require remote employees to use a VPN when accessing corporate resources. However, it should not specify a particular VPN vendor, allowing flexibility to change vendors based on technological and compliance requirements.

System-specific Policy

The third type of policy is system-specific. System-specific policies outline the recommended security configurations for a particular system or network. For example, the firewall policy defines the rules and security configurations when inspecting incoming and outgoing network traffic.

What You Should Consider when Developing a Security Policy?

There is no standard security policy that fits all organizations. Polices will be different according to each organization's work, its processes, and the type of technological solutions it uses. The workforce distribution (onsite, hybrid, or remote) will also be important when crafting your security policy.

Different areas should be considered when organizations create their security policies.

- **Organization security objectives**: Specify the main organization objectives for creating security policies. What do they aim to achieve? For example, the main objectives of all organizations are maintaining the confidentiality, integrity, and availability of data and information assets. Based on this answer, the security policies will be crafted.

- **Distributed architecture**: Today's organizations' IT infrastructure spans on-premise and cloud infrastructure. Organizations should consider how their applications and data are distributed and decide the best security measures to protect them. For instance, distributed networks suffer from more security vulnerabilities than local architecture, which necessitates more security controls and best security practices to counter these risks.

- **Data classifications**: Before creating any policy, data should be classified according to their sensitivity and criticality to organization work. For instance, sensitive data, such as regulated information, should be assigned high-security controls that should be reflected in the policy.

- **Compliance requirements**: Specify the compliance requirements to which the organization is subject. For example, if the organization handles sensitive customers' financial payment information, it could be subject to PCI DSS. Handling patient's information will make the organization subject to HIPAA regulation. Each regulatory body will impose specific protection requirements, which vary from one to one. However, the standard protection requirement begins with installing a firewall, governing user access to sensitive information (by using IAM solution, for example), using encryption, and installing antivirus and antimalware.

- **Regular updates**: Technological advancements are constant and ongoing, so cyber risks continue to evolve. It is crucial to update your security policies to address these evolving threats regularly. Make sure your organization is well-informed about the latest cyber threats and the most effective technological and management solutions available to protect sensitive data and other IT assets. Incorporating this awareness into your existing security policies is essential.

There are policy frameworks published by official bodies, such as the NIST that your organization can benefit from when creating its security policy. The following are the most common ones.

- The NIST cybersecurity framework (https://www.nist.gov/cyberframework)
- ISO/IEC 27001 (https://www.iso.org/standard/27001)

- The European Union Directive 2022/2555 (https://www.enisa.europa.eu/topics/cybersecurity-policy/nis-directive-new)

While numerous frameworks and guides are available, the three mentioned above are widely implemented worldwide. In addition to those already mentioned, various industry bodies and regulatory standards have published their own frameworks.

Important Security Policies for Endpoint Devices

In this section, we will cover the main security policies related to securing endpoint devices.

BYOD Policy

BYOD is a security policy that organizations use to regulate how employees, third-party contractors, and anyone who has access to organizational IT assets and data can use their own computing devices (e.g., smartphone, tablet, and laptop) for work-related activities.

Employees can use their devices for various tasks, such as checking their work emails, accessing cloud applications, accessing organization sensitive files, and other work tasks.

The BYOD security will define the acceptable use of employees' personal devices for work activities. It will also explain the employer's rights when accessing employee-owned devices and what level of access employers will have over their employees' devices.

Each organization will have its BYOD based on its work and needs. However, a typical BYOD policy will contain the following elements:

- **Allowed device types**: The BYOD policy should first define the types of devices allowed to be used at work. For example, laptops, smartphones, tablets, home workstation, and their tech specifications – such as OSs and their support to FDE.
- **Define needed security requirements**: Define the technical security solutions and protective measures that should be installed and adopted on endpoint devices to protect them from cyberattacks, such as antivirus, encryption solutions, and creating complex passwords to protect access to the OS, email client, and other work-related applications. The policy should also define how endpoint device OSs and applications should often updated and patched with the latest security fixes.
- **Define acceptable use**: The policy should clearly define what is acceptable and not acceptable when using personal devices for work. For example, the policy should specify that employees cannot install free internet programs and download programs from pirated websites, such as Torrent, into devices used for work purposes.

- **Define data protection requirements**: This section will list different techniques used by organizations to secure sensitive work data, such as using VPNs to access remote corporate resources and encrypting data stored in endpoint devices.
- **Remote wiping**: The organization may request permission to access employee devices remotely to wipe work-related data (for example, stored in a specific folder). If this issue is required, the policy must clearly define it.
- **Compliance monitoring**: If the employees are processing or storing sensitive customer data. Then, the employer may request the right to access employee-owned devices and inspect them for auditing purposes.
- **Define responsibilities**: BYOD should clearly define employees' and employers' obligations when using personal devices at work. This includes seeking technical assistance from the company's IT department. The policy should also include a section about reimbursement and expenses to define who is responsible if the device needs maintenance and who will pay for data plans.

USB Policy

A USB policy will define the rules governing the usage of USB devices at work. This includes company-owned and employee devices used for work-related purposes.

The USB usage policy aims to manage security risks associated with connecting external storage devices to endpoint devices and the company computer network. Here are the main elements of a suggested USB usage policy.

- **Define allowed USB devices**: The first thing should be to define the types of USB devices permitted to connect to company endpoint devices. For example, external hard drives, USB sticks, USB mice and keyboards, USB speakers, printers, and any device that can connect via a USB port. Most companies do not allow employees' private USB devices to be connected to company resources.
- **Register USB devices**: A policy can require a USB device to be registered first (using its hardware serial number or the MAC address if it has one) before it can be attached to company devices.
- **Define security controls when using USB devices**: Specify the security checks needed when connecting USB devices to the endpoint device. For example, scanning USB devices using a dedicated anti-malware solution.
- **Physical protection of USB devices**: As we know, USB devices, especially those used for data storage, are portable, which makes them susceptible to physical theft. To protect data stored on removable

storage devices, the USB policy should require encryption to protect USB data and/or protect the USB device with a password if applicable.

VPN Access Policy

The VPN policy governs how remote users can access corporate resources via insecure connections – such as the internet. Here are the main elements of a VPN policy:

- **Define user groups**: A VPN policy must first specify the users required to use a VPN service. This can be done by checking each user's job role, as some users may not require VPN access.
- **Specific computing device technical requirements**: Some VPN services need special technical requirements to run. For example, a VPN client software may not run on Android devices while supporting desktop Windows, Linux, or macOS OSs.
- **Authentication method**: Specify the authentication mechanism for accessing the VPN service – such as using 2FA and MFA.
- **VPN client**: List the VPN applications that can be used as a client to connect to the organization's VPN gateway.
- **Encryption algorithms**: Specify the encryption algorithms used to secure the connection – such as AES, RSA, or TLS 1.2+.
- **VPN usage logging**: The policy may require logging VPN usage for auditing and compliance purposes.
- **Usage policy**: Define the acceptable usage policy during VPN usage. For example, prevent circumventing web restrictions imposed by ISPs and governments to visit restricted websites (e.g., Torrent websites) when using a VPN service.

End-users Security Awareness Training

Humans remain the weakest link in any cyber defense strategy, regardless of the security solutions installed on both endpoint devices and network perimeters. A single error caused by careless employees may introduce devastating malware (e.g., ransomware) into their organization's IT environment. This is why educating employees about cyberattacks and mitigation strategies remains a top priority to strengthen organization cyber defenses.

Any employee cybersecurity defense strategy must contain the following three elements:

- Email security
- Malware recognition and mitigation strategies
- Password security

Email Security

Phishing and other types of SE attacks are among the top attack vectors hackers use to infiltrate organizations' IT environments. Educating users about how phishing attacks work practically (via phishing simulators) is essential to mitigate such attacks.

Other areas include teaching users how to open email attachments securely and general best practices for using email securely – such as knowing how to spot phishing emails by reading the email content.

Malware Recognition

Users should understand the different methods that threat actors employ to infect people with malware. This knowledge is essential for employees to mitigate malware attacks, specifically ransomware prevention education.

The cybersecurity awareness for employees should also include:

- Educating employees about the latest cyber threats and how to mitigate them
- Teach employees how to report suspicious activities when they encounter them

Employee training could be in traditional classrooms or via online lessons. The live online lessons are better because it allow repeating tutorials and are more convenient for remote workers.

Password Security

Password security is critical because most organizations still use passwords as the primary authentication method to protect their digital resources. Passwords are also essential to secure endpoint devices and the numerous cloud applications used today by remote employees.

Here are the main security precaution measures for keeping passwords secure:

- Employees should not share their passwords with others – or send them via insecure channels like WhatsApp or Facebook Messenger.
- Do not use the same passwords to protect different online accounts
- Do not write passwords on a piece of paper, like sticky notes, and paste them on insecure locations – like your computer screen.
- Employees should use a password manager to help them create and store their passwords.
- Employees should learn to avoid sharing their passwords with anyone – including their organization's tech support staff.

Summary

Endpoint security is a crucial component of an organization's overall security strategy that requires IT professionals' focused attention. The risks associated with endpoint devices are varied, and a successful security breach can severely impact an organization's reputation and operations. IT professionals must establish a robust endpoint security strategy comprising proactive security measures, comprehensive employee training, and continuous endpoint monitoring. By implementing these measures, organizations can effectively mitigate endpoint security risks and safeguard sensitive information against various cyber threats.

In the next chapter, we will continue our discussion on protecting digital assets, focusing on enterprise defense measures.

Notes

1 Adaptive, "New Research Shows 50% of Endpoints are at Risk", Accessed 2025-04-01. https://adaptiva.com/resources/report/managing-risks-and-costs-at-the-edge

2 IBM, "Cost of a Data Breach 2022", Accessed 2025-04-02. https://www.ibm.com/reports/data-breach

3 Morphisec, "The Third Annual Study on the State of Endpoint Security Risk", Accessed 2025-04-02. https://www.morphisec.com/hubfs/2020%20State%20of%20Endpoint%20Security%20Final.pdf

4 Financesonline, "Number of Internet of Things (IoT) Connected Devices Worldwide 2022/2023: Breakdowns, Growth & Predictions", Accessed 2025-04-02. https://financesonline.com/number-of-internet-of-things-connected-devices

5 Adaptive, "New Research Shows 50% of Endpoints are at Risk", Accessed 2025-04-02. https://adaptiva.com/resources/report/managing-risks-and-costs-at-the-edge

6 Upwork, "Upwork Study Finds 22% of American Workforce Will be Remote by 2025", Accessed 2025-04-02. https://www.upwork.com/press/releases/upwork-study-finds-22-of-american-workforce-will-be-remote-by-2025

7 Expertinsights, "50 Endpoint Security Stats You Should Know in 2023", Accessed 2025-04-02. https://expertinsights.com/insights/50-endpoint-security-stats-you-should-know

8 Github, "Cain & Abel Github Repo", Accessed 2025-04-02. https://github.com/xchwarze/Cain

9 Openwal, "John the Ripper Password Cracker", Accessed 2025-04-02. https://www.openwall.com/john

10 Pcidssguide, "What Are the Firewall Requirements for PCI DSS?", Accessed 2025-04-02. https://www.pcidssguide.com/pci-dss-firewall-requirements

11 IBM, "Cost of a Data Breach 2022: A Million-Dollar Race to Detect and Respond", Accessed 2025-04-01. https://www.ibm.com/reports/data-breach

12 Verizon, "2021 Data Breach Investigations Report (DBIR)", Accessed 2025-04-02. https://www.verizon.com/business/resources/reports/2021-data-breach-investigations-report.pdfx

13 Eftsure, "36 Phishing Statistics in 2022: Don't Take the Bait!", Accessed 2025-04-02. https://eftsure.com/statistics/phishing-statistics-2022

14 Getastra, "81 Phishing Attack Statistics 2023: The Ultimate Insight", Accessed 2025-04-02. https://www.getastra.com/blog/security-audit/phishing-attack-statistics

15 Verizon, "Data Breach Investigations Report 2022: Summary of Findings", Accessed 2025-01-22. https://www.verizon.com/business/resources/reports/dbir/2022/summary-of-findings

16 Microsoft, "Security Signals March 2021", Accessed 2025-04-02. https://www.microsoft.com/en-us/security/blog/2021/03/30/new-security-signals-study-shows-firmware-attacks-on-the-rise-heres-how-microsoft-is-working-to-help-eliminate-this-entire-class-of-threats

17 NIST, "NIST Special Publication 800-12 Revision 1", https://nvlpubs.nist.gov/nistpubs/SpecialPublications/NIST.SP.800-1s2r1.pdf

10

Enterprise Network Defense Strategies

In today's information age, most organizations rely heavily on digital assets – such as information, data, and other digital technologies – for daily functions. This significant dependence on IT systems causes any sudden disruption can potentially result in catastrophic consequences for the affected company. To mitigate such risks and ensure the optimal performance of these technological solutions, it is imperative to have a robust cybersecurity strategy in place. This comprehensive strategy must address all potential cyber threats and propose effective defense measures to thwart them, safeguarding the organization's digital infrastructure and ensuring business continuity.

Developing proper security defenses for any organization requires costly resources. However, the cost of not having an adequate security defense will result in substantial financial and reputation losses. For instance, according to an IBM report titled "Cost of a Data Breach Report 2023", the global average data breach cost in 2023 was USD 4.45 million. This number has increased by about 15% over three years.

The lack of implementation of a cybersecurity defense plan does not stop on financial loss alone. For instance, a cybersecurity strategy will allow organizations to exploit the numerous advantages driven by technology, increase competitive advantage, and enhance customer and stakeholder trust.

In this chapter, we will continue our discussions on the best methods to implement cyber defense controls and strategies to protect organizations' IT environments. After finishing the endpoint device area in the previous chapter, we are focusing on technical and procedural controls in an organizational context.

Network Perimeter Defense Components

When considering implementing cyber defenses in an enterprise context, we must first implement proper defenses on the gateways or organization network perimeter.

A network perimeter is similar to a castle; it ensures everything inside the castle is protected from outside dangers. Of course, a castle will have doors that allow entrance and exit. In IT security, these doors are called network gateways, and they facilitate interconnecting the organization's internal resources (such as the intranet) to the outside world – the internet.

DOI: 10.1201/9781003008279-10

A network perimeter can be either a hardware or software solution that works to prevent external malicious activities from entering the network. Under this definition, a network perimeter includes all networking devices, such as routers and switches, and security solutions, such as firewalls, intrusion detection systems (IDS), intrusion prevention systems (IPS), and virtual private networks (VPNs), used to keep network operations safe.

In the following lines, we will discuss the function of each component and how it helps strengthen the overall network security defense.

Network perimeter is the first defense against external cyber threats targeting enterprise networks. A typical network perimeter includes the following components.

Firewall

In the previous chapter, we discussed personal firewalls and how installing one on all endpoint devices (if applicable) is important.

A network perimeter firewall is a more robust solution to filter traffic on network gateways. In terms of functionality, personal and network firewalls filter incoming and outgoing traffic using predefined security rules. However, network firewalls scan the entire network traffic and have different types.

A network firewall can come in different forms:

1. A hardware device
2. A software solution
3. Software-as-a-service (SaaS) solution managed by external providers
4. A public cloud solution
5. A private cloud (virtual)

There are different types of network firewalls. Organizations select the best solution based on their data and IT infrastructure needs.

Firewalls Types

There are different types of firewalls, depending on their functions.

Proxy Firewall

A proxy firewall operates at the application layer of the Open Systems Interconnection (OSI) model and acts as a gateway between internal users and the internet. For instance, a company may set a proxy firewall in its computer network to filter its employees' access to internet resources. The proxy firewall may contain a blocklist of websites employees are not allowed to visit during working hours, such as Facebook and Instagram or online streaming services like Netflix and Hulu. The proxy firewall will block the connection if employees try to access one of these restricted sites.

Proxy firewalls can also be configured to prevent users from using some applications within the corporate network, such as internet messaging apps like WhatsApp and Facebook Messenger.

Proxy firewalls can provide additional security features, such as the following:

- Content filtering – prevents users from accessing malicious websites such as phishing and torrent websites.

- Scan for malware – check for files downloaded from the internet and quarantine them if they contain malicious code before they access the corporate network.

- Data leak prevention – through scanning email contents and attachments to check if someone is trying to exfiltrate sensitive business information outside the company.

- Conceal sender IP address – a proxy firewall will have its IP address. This is a major security benefit because external destinations will never see the actual IP address of the internal sender.

Stateful Inspection Firewall

A stateful inspection firewall (also known as a dynamic packet filtering firewall) is located at layers 3 and 4 of the OSI model. It scans the currently active network connections (both incoming and outgoing network communications), looking for malicious traffic or other risk signals. Stateful firewalls filter network traffic based on each network connection's state and context.

State: The firewall records each active connection's state as specified in the session packets. It tracks the entire lifecycle of a connection, from establishment to termination, and maintains a table of active sessions.

Context: The connection context refers to information that can be used to determine repeated patterns. For instance, it includes the source and destination IP addresses, connection ports, protocols, and other metadata information.

Stateful firewalls can inspect data inside network packets to determine whether they contain malicious code or behavior that can compromise network security. They can also track complex protocols that use multiple connections, such as FTP.

The way a stateful firewall inspects network traffic makes it suitable to enforce network security in different ways:

- Organizations can use it to enforce security policies across the entire network environment. This provides a centralized solution to govern users' access to network-protected resources.

- It can intelligently integrate artificial intelligence (AI) and machine learning (ML) technologies to filter network traffic. This will effectively reduce network administrator time and enhance the overall network security.
- It can be used to halt advanced attacks such as brute force, denial of service (DoS), malware attacks, and data leaks because of its ability to perform deep packet inspection of network connections.
- Can filter network connections based on the ports computing devices use to access protected resources.
- It provides better performance and scalability than traditional packet filtering firewalls because it only inspects the initial packet of a connection and then caches the connection details.

These features allow stateful inspection firewalls to offer advanced protection against various network threats while minimizing the impact on network performance.

UTM (Unified Threat Management)

As the threat landscape continuously evolves, installing multiple solutions to protect the network perimeter against each threat becomes daunting and costly for organizations. For instance, to adequately protect your organization's network, you need to install at least the following solutions: IPS, IDS, a stateful firewall, and antivirus to handle different cyber-attacks.

UTM evolved as a way to combine multiple network security solutions into one device. For instance, a typical UTM will provide the following functions:

- Stateful inspection firewall
- IPS
- Antivirus
- Content filtering – email and web filtering features
- Data loss prevention
- VPN

UTM provides a unified solution for cost-effectively handling different security aspects of network security, and it is well-suited for small and medium-sized companies with limited IT infrastructure and budget. It simplifies deployment and management by combining multiple security functions into a single solution.

However, UTM solutions often provide limited configuration capabilities when you want to customize each security function individually. The lack of customization encouraged manufacturers to develop a new type of advanced firewall, the Next-Generation Firewall (NGFW).

Next-generation Firewall

The next-generation firewall (NGFW) emerged as an evolution of the UTM concept, providing numerous advanced security features, such as:

- Standard firewall capability – Deep packet inspection.
- IPS capabilities to detect and prevent known and emerging threats, including malware, exploit kits, and other security vulnerabilities.
- SSL/TLS inspection – NGFWs can decrypt encrypted traffic to reveal suspicious connections hidden in encrypted channels.
- NGFWs provide improved access controls over network resources. For instance, they can identify and apply policies based on user identities, applications, and content.
- Some NGFW solutions include sandboxing and other advanced techniques to detect and mitigate advanced malware and zero-day threats. This gives advanced protection against malware attacks.
- NGFWs can integrate with other security technologies, such as security information and event management (SIEM) systems, which provide comprehensive network visibility.
- Some NGFWs can connect to threat intelligence feeds to better fight against zero-day attacks.

Boarder Router

Border routers are networking devices used to forward network packets between autonomous systems. For example, they can connect internal network resources with the outside world, such as the internet. These devices play a crucial role in network infrastructure by interfacing between different network domains.

Because of its ability to filter incoming and outgoing traffic, a border router can be used to enforce different security policies and thus strengthen the entire network's security. This filtering capability allows border routers to provide several key security functions:

- Implementing Access Control Lists (ACLs) allows a network admin to filter traffic based on predefined rules. Border routers can also segment networks, improve overall network security, and reduce the effectiveness of data breaches.
- Provide the basic firewall functionality to block unauthorized access attempts.
- Provide a Network Address Translation (NAT) feature, which hides internal IP addresses from external networks.

- Can terminate unauthorized VPN connections to provide secure remote access.
- It can help fight DDoS attacks and protect against distributed denial-of-service (DDoS) attacks.

Intrusion Detection Systems

An Intrusion Detection System (IDS) is a security solution that comes as a software tool installed on endpoint devices, installed on networking servers as a program, a cloud service or as a dedicated hardware appliance installed on a network perimeter. IDS monitors network traffic or other devices for suspicious activities or security policy violations. When the IDS detects malicious behavior, it can either generate alerts or take automated actions to cease potential security threats.

IDS employs various detection methods, including signature-based, anomaly-based, and hybrid. For instance, the signature-based method compares observed behavior against known attack patterns, while anomaly-based detection identifies deviations from normal system or network behavior. The hybrid IDS type combines the two detection methods to avoid the limitations of signature and anomaly detection methods. The signature database of the signature-based IDS solution must be regularly updated with new threat intelligence to counter emerging threats.

IDS solutions can be deployed in diverse environments, from small businesses to large organizations. For instance, an online bank might implement an IDS to monitor traffic to its online customer-facing dashboard for signs of fraudulent activities or attempts to exploit any security vulnerabilities.

It is worth noting that IDS cannot stop cyberattacks alone. For instance, IDS solutions are often integrated with other security solutions, such as a security information and event management (SIEM) system, which receives real-time alerts from IDS systems (and other systems) across the network to identify and respond to discovered cyberattacks.

Installing an IDS solution will strengthen your network defenses and help your organization comply with regulatory compliance regulations, such as the Payment Card Industry Data Security Standard (PCI-DSS), which mandates having an IDS solution to detect and/or prevent intrusions into the network as stated in its official guide "Requirement 11: Regularly test security systems and processes".[1]

There are many IDS solutions in the market out there; the following are three free solutions:

1. Snort[2]: An open-source, widely-used network intrusion detection system that applies a set of rules to identify malicious network activity, scanning packets for matches and generating alerts for users when suspicious activity is detected.

2. OSSEC[3]: An open-source host-based intrusion detection system that can be customized to meet your security needs by configuring it extensively, adding custom alert rules, and scripting automated responses when alerts are triggered.

3. Suricata[4]: Another open-source network analysis and threat detection software. It can integrate with different commercial and open-source security solutions.

Intrusion Prevention Systems

An Intrusion Prevention Systems (IPS) is a network security solution that detects and responds automatically to security events without human intervention. IPS commonly comes integrated with IDS functionality. For instance, an IDS is a reactive solution that cannot stop security threats alone. In contrast, IPS solutions play a proactive role by detecting and acting immediately to prevent discovered cyber-attacks from escalating.

IPS monitors network traffic in real time; once detecting malicious behavior or other attack patterns, it will automatically record the attacker's IP address and initiate an automated response that is already configured by the network admin. The IPS will then send a report to the network admin describing what happened and block all future connections from the attacker's IP address.

There are five types of IPS solutions:

1. **Network-based intrusion prevention system (NIPS):** Monitor the entire network traffic for malicious activities and abnormal patterns. NIPS are typically deployed at network entry points to inspect all incoming and outgoing traffic. They can detect and block different types of threats, such as known exploits and malware.

2. **Wireless intrusion prevention system (WIPS):** Monitor the wireless network (frequency analysis) for malicious activities. WIPS is considered critical for securing Wi-Fi networks against unauthorized access, rogue access points, and other types of wireless attacks. Organizations also use WIPS to enforce wireless security policies and ensure compliance with regulatory standards.

3. **Host-based intrusion prevention system (HIPS):** A software program installed on endpoint devices (if applicable, as we cannot install IPS agents on many Internet of Things (IoT) device types), used to monitor traffic coming or leaving the host for any malicious activities. HIPS can be used to detect changes in the configuration settings of the host that may raise a red flag, as well as to execute packet inspection to detect abnormal connections and terminate them instantly. HIPS also often includes additional security features like application control, memory protection, and file integrity monitoring.

4. **Network behavior analysis (NBA):** This is also known as "Behavior Monitoring". NBA works by analyzing internal network traffic to detect abnormal activities such as DDoS, malware, security policy violations, and anything that points to suspicious activity. NBA uses sensors (network analyzers) or gets network traffic from routers, switches, or other networking devices to inspect it. Modern NBA solutions detect malicious activities using ML technology. When the NBA is installed on a computer network for a long time, it will learn the normal network activities, enabling it to detect sudden changes in network traffic. This adaptive learning capability allows NBA systems to identify zero-day threats and sophisticated attacks that might evade traditional security solutions relying on signature-based techniques.

5. **Cloud-based intrusion prevention system:** The increased adoption of cloud services has encouraged many companies to launch cloud-based IPS solutions on a subscription basis. These systems protect cloud infrastructure and applications by monitoring traffic between cloud instances and on-premises networks.

Virtual Private Networks

The VPN is a technology that establishes a secure connection between two points over an insecure medium such as the internet. VPN is commonly used to secure access to remote resources. For example, during the COVID-19 pandemic, organizations widely used VPN technology to connect remote employees to corporate internal networks.

The encrypted tunnel established by the VPN helps ensure all sensitive data passing through the connection is protected from eavesdropping or other types of attacks, such as man-in-the-middle attacks. This encryption process typically uses robust protocols like IPsec or SSL/TLS to safeguard data in transit.

VPN technology is widely used in enterprise environments to secure access to their sensitive resources, including cloud instances, internal databases, and proprietary applications. Offsite employees can use their portable devices, such as smartphones, tablets, and laptops, to access remote resources securely, regardless of their physical location.

These were the primary components used to secure organizations' computer networks. It is worth noting that today's IT environments are scattered across different geographical areas and utilize both on-premise and cloud infrastructure. Such a setting is known as a hybrid environment and necessitates additional security measures to safeguard data and programs. While this complicates the process of securing these resources, IPS/IDS and VPN can still be used extensively to secure connections between cloud instances and the organization's on-premises resources.

In addition to the solutions mentioned to secure network perimeters, organizations operating in hybrid environments can use the following solutions to improve their cloud security:

1. Cloud Access Security Broker (CASB): This software program can be installed on-premise or in the cloud. It sits between cloud-based applications and their users and monitors all interactions between them for any sign of malicious activities. CASBs also provide visibility into cloud application usage and can enforce security policies.

2. Cloud Security Posture Management (CSPM): This solution helps organizations identify and remediate risks across their cloud infrastructure. It provides continuous monitoring of cloud configurations and compliance with security best practices.

3. Multi-factor Authentication (MFA): MFA, including using at least one authentication factor that is dependent on biometrics, is essential to protect both on-premises and cloud-based resources. It adds an extra layer of security beyond just passwords.

4. Data Loss Prevention (DLP): A security solution used to prevent internal users from sharing sensitive information and files outside the organization's IT environment. DLP solutions can prevent unauthorized disclosure of information across on-premises systems, cloud-based locations, and endpoint devices. A DLP solution must be installed to comply with HIPAA and GDPR.

5. Security Information and Event Management (SIEM): SIEM aggregates threat data from different data sources across the organization's IT environment and provides complete visibility over all resources used across the network. It helps in the real-time analysis of security alerts generated by applications and network hardware.

6. Cloud-native security tools: Depending on your cloud hosting provider, many cloud providers offer native security tools that integrate by default with their platforms, such as AWS GuardDuty or Azure Security Center, to protect cloud customers' resources. These tools often provide features like threat detection, compliance monitoring, and vulnerability testing.

7. Container security: For organizations using containerized applications, specialized container security tools are used to secure the entire container lifecycle, from its build to deployment and run.

8. API security: Leveraging cloud resources necessitates using APIs to connect with various cloud services. API security solutions should be employed to protect these "connection points" from malicious exploits.

Network Sandboxing

Sandboxing is a security technology that uses an isolated environment to test suspicious code before executing it in a production environment or gaining access to the internal company network.

Sandboxing allows the security team to safely run and analyze any suspicious code, application, or digital artifacts in a closed environment to inspect their malicious behavior.

Sandboxing technology was proposed to solve a key security issue: zero-day attacks or threats. A zero-day threat is an unknown security vulnerability in your software or hardware assets discovered and exploited by threat actors. Part of its name, "zero-day", originates from the fact that such vulnerabilities were just found by the exposed system vendor or before they knew about them. By using sandboxing technology, suspicious artifacts could be executed within the sandbox without posing any security risks to your system or platform.

It is worth noting that sandboxing technology is not merely used for security purposes. For instance, development teams use sandboxing to integrate different parts of software projects and see how they run together before releasing them to the production environment.

Sandboxing Types

There are different types of sandboxing according to their purpose.

Application Sandbox

This approach isolates an application, such as a mobile app, from an untrusted source (e.g., unknown developers or websites), from interacting with the underlying system. It protects the application from external threats, like malware or hackers, and ensures that users cannot inadvertently interact with the underlying platform while using the sandboxed application.

Web Browser Sandbox

This approach executes web browsers within an isolated environment that prevents interaction with the underlying system. This isolation effectively shields the system from web browsers and web-based email threats, such as malware from malicious or compromised websites and harmful attachments in web emails. This provides robust protection for enterprise environments against similar threats.

We can differentiate between two main web browser isolation techniques:

1. Locally installed isolated browser: In this approach, the web browser is installed within a controlled environment on the user's computing device. Commonly, it is installed within a virtual machine (VM) instance or a container. This approach guarantees all web-based

threats, such as malware, will not infect the host operating system. While this approach is convenient for individual users, it introduces some technical challenges in the enterprise context, such as:

a. Running a virtual machine on each end-user device is resource-intensive and may not be feasible for old computing devices.

b. Special configurations should be made for each end-user device, which will require additional work for the IT support staff.

2. Remote browser isolation: In this approach, the web browser runs in an isolated environment hosted on the cloud or a specific server within the organization's network. The user interacts with an instance of the browser while the actual execution occurs remotely. While this approach gives organizations many advantages, such as being easy to manage and not requiring intensive computing power on the end user device, the main drawback of this approach is that users may expect some latency while browsing the web, especially if the network bandwidth is limited.

Security Sandboxing

In a security sandbox, an organization dedicates an isolated environment to inspect suspicious code, files, and other artifacts before allowing them to enter its IT environment. This approach provides enhanced protection against zero-day exploits and vulnerabilities, in addition to scanning email attachments and files uploaded from applications for malware (ransomware, trojans, spyware, worms).

Network Segmentation

Network segmentation (see Figure 10.1), also known as Network Segregation or Network Partitioning, enhances computer network security and performance through isolation. Segmentation allows an organization to divide its computer network into pieces, and each piece can have its own configurations, ultimately maximizing performance, security, and organization.

Segmentation helps organizations automatically decide what traffic can access a specific segment, which helps achieve the principle of least privilege.

Segmentation can be achieved using two primary methods:

- **Physical segmentation:** This is achieved by isolating network segments using routers, switches, or firewalls. A firewall is commonly installed to filter all traffic coming or leaving each segment. This option is costly because each segment requires its own physical infrastructure, such as cables and networking devices.

- **Logical segmentation:** This method is widely used to segment networks into manageable chunks. It is a cost-effective solution as it does not require purchasing networking devices. Logical segmentation is achieved using virtual local area networks (VLANs) or software-defined networking (SDN) to create separate network segments within the same physical infrastructure.

Segmenting a network into isolated pieces helps achieve numerous benefits for organizations:

- Reduce network congestion. For example, in a bank system, we can set the public bank website on one network segment and the online banking application used to provide services to bank customers in an isolated segment. This ensures the heavy traffic of users' regular web browsing activities does not affect the online banking app.

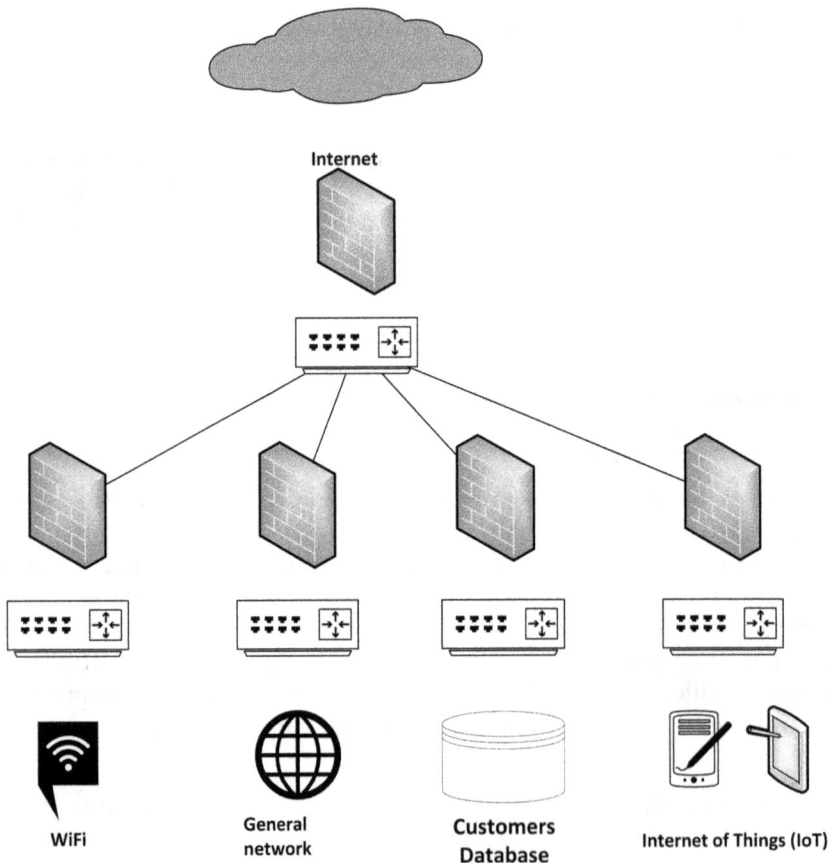

FIGURE 10.1
Network segmentation.

- Segmentation improves cybersecurity defenses. For example, if adversaries sneak in and gain access to the target network, their ability to spread remains limited to the infected segment only. The same applies to malware infection; malware can infect only the resources within the infected segment.

- Segmentation allows an organization to put its most valuable assets, such as customers' personally identifiable information (PII) and financial/medical records, in one segment and isolate it from the internet, which greatly enhances its protection from cyber-attacks.

- Segmentation helps secure devices that are not protected enough within an organization's IT environment. For example, many IoT devices cannot accept antivirus or antimalware installations due to their low computing power. Grouping these devices in one network segment can improve their resilience to cyberattacks by installing advanced security defenses on the IoT network segment.

- Segmentation helps reduce costs associated with compliance regulations. For example, the system used to process patients' payments can be settled in a dedicated segment. Only this segment, not the entire network, will be subject to the PCI DSS regulation.

- Enhance incident response: Segmentation can be very useful for incident response. For instance, segmentation prevents the spreading of the infection to the network areas and speeds up the process of containing incidents. During incidents, security teams can also locate the affected area quickly, which greatly enhances the overall incident response activities.

- Zero-trust architecture: Segmentation is considered a critical component of implementing a Zero-Trust security model, where Trust is never assumed for both users and systems, and everything should be verified continually before accessing sensitive data or using applications.

Zero Trust Architecture

Zero Trust Architecture (ZTA) is increasingly becoming a cornerstone of modern cybersecurity defense strategies, especially for organizations operating in hybrid and cloud environments. This model moves away from the traditional "trust but verify" scheme, which inherently assumes that all entities within the network perimeter are trustworthy by default, to a more rigorous "never trust, always verify" approach. In an era where cyber threats continue to be very sophisticated and evolve rapidly, Zero Trust provides a robust framework to mitigate risks by ensuring that no user, device, application, or service is trusted by default, regardless of their location within or outside the network.

Zero Trust Key Components

At its core, Zero Trust is built on several key principles designed to enhance security posture:

- Least privilege: This principle ensures that users and devices are granted the minimum level of access necessary to perform their job duties and nothing else.
- Micro-segmentation: This model divides the network into smaller, isolated segments, and each segment operates independently. This ensures that even if one segment is compromised, the breach does not spread across the entire IT environment.
- Continuous verification: Unlike traditional models that authenticate users only at the point of entry (mostly on network perimeter), Zero Trust requires ongoing verification of user, device, and app identities

How Zero Trust Complements Network Segmentation and Other Defense Mechanisms

Zero Trust does not operate in isolation; it enhances and integrates with existing security measures such as network segmentation, firewalls, and IDS systems. While network segmentation divides the network into segments, Zero Trust takes this a step further by enforcing strict access controls within each segment. This layered approach ensures that even if an attacker bypasses one layer of defense, they are still restricted by the granular controls of Zero Trust.

Practical Steps for Implementing Zero Trust in an Enterprise Environment

To implement zero Trust in your enterprise IT environment, you can follow the following five phases:

1. Assess the current environment: Begin by identifying which assets, users, workflows, and critical data require protection
2. Define access policies: Establish clear policies based on the principle of least privilege to ensure access is only granted to those who need it
3. Deploy Identity and Access Management (IAM) Solution: Implement multi-factor authentication (MFA) and single sign-on (SSO) to strengthen identity verification and store all access credentials along with their permission in a dedicated IAM solution to simplify credentials management

4. Leverage micro-segmentation: Use software-defined networking (SDN) or other technologies to create micro-network segments

5. Monitor and adapt: Continuously monitor user behavior and network activity to detect anomalies and adjust access controls as needed

Examples of Zero Trust Solutions

Several organizations have developed frameworks and solutions to help enterprises adopt the Zero Trust model. Here are the most prominent ones:

- Google BeyondCorp[5]: BeyondCorp represents Google's practical implementation of the Zero Trust model, rooted in over a decade of internal experience and enriched by insights and best practices from the broader cybersecurity community. By redefining access controls – moving them away from the traditional network perimeter and focusing instead on individual users and devices – BeyondCorp allows employees to work securely from virtually any location. This approach eliminates the reliance on conventional VPNs, offering a more flexible and robust solution for modern, distributed work environments.
- Microsoft Zero Trust Framework[6]: Microsoft's approach integrates Zero Trust principles across its ecosystem, including Azure Active Directory, Microsoft 365, and endpoint security solutions. It emphasizes continuous verification and least privilege access across all layers of the IT environment.

Penetration Testing

Penetration testing, or pen testing, is a security exercise in which cybersecurity professionals try to exploit vulnerabilities in target computing systems and networks. Pen testers use the same techniques and tools black hat hackers use to exploit networks. However, they do this for a good reason: to discover and fix security vulnerabilities before threat actors exploit them.

Pen testing safeguards enterprise computer networks for these critical reasons:

- Pen testing replicates genuine attack scenarios through stealth assessment. Security teams conduct unannounced attacks against network infrastructure to evaluate security control effectiveness under real-world conditions.

- Risk prioritization becomes data-driven through pen testing outputs. Organizations can sequence remediation efforts based on vulnerability severity scores, optimizing resource allocation for maximum security impact.
- The testing process exposes architectural flaws, implementation gaps, and configuration/deployment errors in software systems. Development teams can leverage these insights to enhance code security, patch management, and system hardening procedures.
- External third-party pen testing delivers an unbiased security posture evaluation. Independent assessors bring fresh perspectives to identify overlooked vulnerabilities within the technology environment.
- Advanced threat simulation enables proactive defense enhancement. By modeling sophisticated attack chains – from social engineering to lateral movement techniques – organizations can validate detection capabilities against emerging threats.
- Regulatory compliance frameworks mandate regular pen testing. Standards like PCI DSS require periodic assessments to validate security controls used to protect sensitive data.
- Post-incident analysis benefits from pen testing data. Security teams can reconstruct attack paths by comparing incident patterns against previously documented testing scenarios, accelerating forensic investigations.

Physical Security

Physical security is concerned with securing the IT infrastructure from any threats that can affect its performance or make it unavailable. For instance, an organization's IT infrastructure, such as networks, servers, endpoint devices, and any electrical equipment needed to run the computer network, is subject to numerous risks, such as natural disasters (flood, earthquake, fire), theft, sabotage, terrorism, or unauthorized access. We have already covered how to secure network perimeters; however, the physical security of networks is equally important when discussing the creation of a comprehensive cyber defense strategy for enterprises.

When developing a plan for securing an organization's IT infrastructure physically, security teams should consider the following three angles:

1. Access controls
2. Surveillance systems
3. Testing

In the following lines, we will cover each one in some detail.

Access Controls

The first line of defense is the buildings that host the IT infrastructure. Access controls to these buildings are installed to ensure that only authorized people can enter. There are different types of physical access controls.

Mechanical

This includes traditional locks that require a key to open, in addition to keycards and mechanical codes.

Keycards are plastic cards that contain an access key stored within a magnetic strip and are used to unlock doors.

Mechanical codes are locks that require a user to input a secret key in a specific combination to unlock the door.

Electronic Access Control

Electronic access control (EAC) systems use digital technologies to restrict access to buildings and other secure spaces. These systems include the following:

- Electronic keycard readers: These are security devices installed on doors or gates. They use RFID or smart card technology to read the information stored on cards and grant access to restricted areas. Some advanced readers can also detect tailgating attempts.
- Biometric scanners: These utilize unique physical characteristics of people, such as fingerprints, facial features, or retinal scans, to identify individuals. Biometric systems offer high security as they are difficult to forge or share.
- PIN code lock: This security device requires users to provide a pre-set personal identification number (PIN) or a numeric code via a keypad to gain access. It could be installed on the door or used as part of a large access control system. Some advanced systems use dynamic PINs that change regularly for enhanced security.

Access control systems have evolved to include more features to strengthen access to restricted resources. This includes:

- Time-based access: This feature can be added to EAC, allowing access only during specific hours or days (for example, from 8:00 a.m. to 3:00 p.m.). This feature is handy for managing contractor access or enforcing work schedules.

- Multi-factor authentication: Combining two or more authentication methods (e.g., keycard, PIN, and facial scanner) for enhanced security significantly reduces the risk of unauthorized access even if one factor (such as keycard or PIN) is compromised.
- Access logs: Most electronic access control mechanisms maintain a detailed record of their operations, such as who entered, when, and where. These logs are crucial for auditing, investigating security incidents, and demonstrating compliance with regulations.
- Remote management: Modern access controls allow administrators to grant or revoke access remotely in real time. This feature is particularly important in emergencies or when managing access for large, distributed organizations.

Surveillance Systems

In the context of securing the physical IT infrastructure, surveillance refers to the act of using different resources, personnel, and technologies to monitor and protect access to physical buildings and other facilities.

The following technologies can be used to implement physical access controls through surveillance:

- Closed-circuit television (CCTV) cameras: Provide real-time monitoring and record movements and access to sensitive facilities, such as server rooms, data centers, and the location of other IT infrastructures, including networking device rooms. CCTV is used to record access to secure facilities and can play a proactive role by preventing criminal behavior. For example, criminals are less likely to attempt breaking into server rooms when they see CCTV everywhere because they know security guards will respond instantly to capture them. Modern CCTV systems come powered with AI technology, often including features like facial recognition, automatic alert systems, and integration with access control databases.
- Motion sensors and alarms: Motion sensors are valuable tools for detecting unauthorized movements when someone tries to access a particular building or room without permission. They should be placed in areas where intruders are most likely to pass, such as doorways, windows, and hallways. Advanced motion sensors can differentiate between human movement and other types of motion, which helps reduce false alarms. These sensors can be integrated with alarm systems to alert security personnel of potential intrusions immediately.
- Environmental systems: The physical risks against IT infrastructure are not limited to human intruders. For instance, a problem in

the air cooling system can lead to fire or prevent some computing equipment from functioning properly. To mitigate such risks, all rooms containing sensitive IT and electrical equipments should be equipped with environmental monitoring systems to measure and alert about any sudden increase or decrease in temperature, humidity, or any change in power supply (decrease or surge). These systems can also monitor for water leaks, smoke, and air quality.

Testing

The testing phase is very important to have a proper physical security plan. It ensures all physical security procedures are working as expected. Audits should be performed at least 3 times a year to ensure optimal security. Here are what security teams should regularly test to ensure proper physical security:

- Physical penetration testing: Organizations should periodically conduct physical penetration testing of their facilities to ensure intruders cannot gain unauthorized access to sensitive rooms and buildings such as data centers and offices. Physical penetration testing is the process of simulating an intruder executing a planned attack to circumvent an organization's physical barriers to gain unauthorized access to infrastructure, buildings, computing systems, and employee offices. This can include testing access control systems, surveillance cameras, and security guard responses.

- Security assessment: Conduct regular physical security audits to assess the organization's immunity to internal and external physical threats. This includes different areas of the property such as entryways, exits, installed physical security solutions, environmental factors surrounding the facility, examining the interior and exterior of the facility, and how visits to the facility are recorded and managed. These audits should also include checking for vulnerabilities in the supply chain and vendor access procedures.

- Test emergency response procedures: In the case of unexpected events, it is critical to ensure the resilience of your organization to different threats such as natural disasters, cyberattacks, or failures in equipment. The emergency response procedures include assessing risks to physical IT infrastructure, identifying critical systems that must have backups, and ensuring emergency systems, such as fire suppression systems, are ready and functioning properly. Emergency procedures should be tested regularly through simulations and practical scenarios to ensure they work correctly. This should also include testing of backup power systems and data recovery procedures.

- Test access control systems: Regularly test all access methods (key-cards, biometrics, PINs) to ensure they function correctly.
- Test environmental monitoring systems: Perform regular tests on temperature, humidity, and electricity monitoring systems to ensure they function correctly and can report anomalies.
- Integration testing: When multiple security systems are integrated (e.g., CCTV with access control), we should perform regular tests to ensure they are working together as intended.

Security Policies and Procedures

In the context of cybersecurity, security policies and procedures refer to the comprehensive set of technical controls and management guidelines that define how an organization should protect its digital assets. These assets include both data and IT infrastructure, such as computing devices and networks. The ultimate goal of these policies is to outline the specific actions required to maintain the confidentiality, integrity, and availability of digital assets. In an enterprise setting, security policies should address the following key areas:

- Access Controls
- Incident Response Management
- Data Handling
- Password Management
- Employee Cybersecurity Training

Access Controls

Access control policies define the rules and technical mechanisms governing how digital asset access is granted, managed, and revoked. These policies are very important to secure access to sensitive data and IT systems as they ensure that only authorized individuals can access sensitive resources based on their roles and responsibilities.

The main elements of access control policies include the following.

Role-based Access Control

Role-based access control (RBAC) assigns permissions based on user roles and job responsibilities. This ensures users have access only to the resources necessary for their tasks – nothing more. For example, an HR employee can access personnel employs records, while an IT staff member can manage network configurations. This approach minimizes the risk of unauthorized access while maintaining operational efficiency across different work environments.

Least Privilege Principle

This approach grants users, applications, and IT systems only the minimum access required to perform their tasks. It reduces the attack surface, limits damage from compromised users accounts, and prevents accidental or intentional misuse of sensitive data.

For instance, under the principle of least privilege (PoLP), if a finance department user's account is compromised by malware, the impact is contained. The malware cannot spread freely across the network since the account has strictly limited access.

Multi-factor Authentication

Multi-factor authentication (MFA) is a critical access control measure for securing sensitive resources. To grant access, it requires additional verification beyond just a password – such as a code sent to a user registered mobile device or a physical security key. This extra layer makes unauthorized access very hard, even if credentials are stolen.

Incident Response

Incident response policies outline the procedures for identifying, managing, and recovering from cybersecurity incidents, such as data breaches, malware infections, or denial-of-service attacks. It also covers incidents caused by non-cyberattacks, such as power failures and internet connection interruptions.

The incident response policies would cover the following main elements.

Incident Identification

Define how to detect and report potential security incidents (e.g., unusual network activity, unauthorized access attempts, or a malware infection). Include severity classification criteria to prioritize response efforts.

Response Plan

Establish a step-by-step plan for containing, investigating, and mitigating incidents. Define clear roles and responsibilities for each team member during the response process. Include containment strategies for various incident types.

Communication Protocols

Specify who should be contacted during an incident (e.g., IT team, management, legal team, or external authorities such as law enforcement). Establish secure communication channels and notification templates for different stakeholders.

Post-incident Review

Conduct a thorough analysis after an incident to identify root causes and improve future response mechanisms. Document lessons learned and implement changes to strengthen security controls based on findings.

Data Handling

Data handling policies define the set of procedures for how an organization manages its data. It defines how data is collected, stored, accessed, processed, and disposed.

The key elements of data handling policies include the following.

Data Collection

Specify the framework on how data should be collected; this includes the legal and ethical boundaries of data collection practice to ensure only the required data to perform work is collected and in a lawful way. For example, many companies are using their customer communications and even voice commands, such as voice commands for Amazon Alexa assistance, to train their ML models. Include explicit consent mechanisms and transparency about collection purposes.

Data Storage

This defines the managerial and security procedures to store data securely to prevent unauthorized access. This includes using encryption to secure data at transit and at rest, securing data storage servers, and executing regular backups to ensure data availability. Implement data classification schemes to determine appropriate storage controls based on sensitivity levels (see Table 10.1).

Data Access

Define who can access stored data and under which conditions. This ensures data remains confidential, prevents unauthorized access to sensitive data, and protects it from misuse, accidental deletion, or modifications. Implement the principle of least privilege and require authentication mechanisms appropriate to data sensitivity.

TABLE 10.1

Data Sensitivity Levels

No	Data Type	Sensitivity Level
1	Social security numbers, financial and medical records, account credentials	High
2	Email address, phone numbers	Moderate
3	All other data that does not require protection	Low

Data Sharing

Define the legal boundaries of sharing data with other parties. This also includes the need for legal agreements to ensure the recipient follows proper data protection standards. Document data transfer methods and maintain sharing logs for accountability and compliance purposes.

Data Disposal

Set guidelines for the safe disposal of data that is no longer needed to ensure that it cannot be recovered again. For example, data destruction algorithms can be used to destroy data on disposal hard drives. Paper documents should be destroyed using a physical shredder machine. Include retention schedules specifying how long different data types should be kept before disposal.

Data destruction tools
  There are many tools to erase data, ensuring it cannot be recovered securely. Here are three tools:

1. DBAN (https://dban.org)
2. BCWipe (https://jetico.com/data-wiping/wipe-files-bcwipe)
3. KillDisk (https://www.killdisk.com/eraser.html)

An example of a data handling policy is requiring all employees to avoid storing sensitive customer data on their workstations. Instead, all sensitive data must be kept on a central, secure server and accessed over the network. When transferring data outside the company network – whether via email, USB stick, or CD – it must be encrypted to prevent unauthorized access.

Public data, such as website content, annual reports, and business filings, has no such restrictions and can be freely shared.

Password Management

Password management defines the rules for creating, storing, and updating passwords to prevent unauthorized access to accounts and IT systems.

A typical password management policy will contain the following key elements:

- *Password complexity*: Create strong passwords with a minimum length (e.g. at least 12 characters), a mix of characters (uppercase, lowercase, numbers, and symbols), and avoid using common words such as pet or spouse name as a part of the password. Proton Pass (https://proton.me/pass/password-generator) is an online open-source password generator tool.

- *Password expiration*: Require regular password changes (e.g., every 60 days)
- *Password storage*: Prevent employees from writing their passwords on papers or keep them unencrypted on their mobile phones. For example, they may be required to use a password manager to keep all passwords stored securely. Psono (https://psono.com) is an example of an open-source self-hosted password manager for companies.
- *Password lockout*: Design your login systems to lockout users after a defined number of failed logins, such as five failed logins after then a user must wait for one hour before trying again. This allows you to halt brute-force attacks executed by automated attack tools.
- *Password uniqueness*: Prohibit password reuse across multiple systems and prevent using previous passwords when making changes.
- *Multi-factor authentication*: Implement additional verification beyond passwords, like one-time codes sent to mobile devices, especially for sensitive systems
- *Password sharing protocols*: Establish guidelines for emergency access and strictly prohibit informal password sharing between staff members.

Employee Cybersecurity Training

Employee training policies ensure that all organization employees understand cybersecurity risks and best practices, which help reduce human error – a major cause of security breaches. For instance, according to the IBM Cost of a Data Breach Report 2024, "IT failures or human error caused nearly half of all breaches, with 19% specifically attributed to phishing attacks".[7] Proper training strengthens an organization's security posture by making employees the first line of defense against cyber threats.

The cybersecurity awareness training of employees should cover the following knowledge areas.

Phishing Awareness

Employees must be trained to recognize and notify about phishing emails, suspicious links, and malicious attachments. Threat actors often disguise emails as legitimate communication from banks, colleagues, or vendors to trick users into clicking malicious links or sharing their account credentials. Regular phishing simulations and real-world examples help employees develop the skills to identify and avoid phishing threats. Training should highlight current phishing tactics such as business email compromise (BEC)[8] and spear phishing[9] targeting executives.

Social Engineering

Social engineering (SE) training educates employees on how threat actors manipulate individuals' minds into revealing sensitive information such as account credentials or other sensitive information. Common SE tactics include impersonation scams, pretexting (creating false scenarios to gain target trust), and baiting techniques, such as leaving infected USB drives in office spaces or parking. Employees should be very careful when sharing information with external parties, even if requests appear to come from trusted sources, such as company managers. Training should emphasize performing verification in response to unusual requests, particularly those involving financial transactions or sensitive data transfers.

Safe Internet Usage

Organizations should provide clear guidelines for employees on safe web browsing, how to download files and programs from the internet, and the secure methods to use company computing devices. For instance, employees must avoid visiting malicious websites such as torrent and pirated content websites, refrain from downloading unauthorized software, and understand the risks of connecting to public Wi-Fi in restaurants, cafes, and public transportation. Training should also reinforce secure password management practices, including complex passphrase creation, regular rotation, and the importance of leveraging MFA to protect user accounts. Password managers should be recommended as a tool to maintain unique credentials across different systems.

Data Handling Procedures

Organizations should establish a clear policy for employees regarding the classification, storage, transfer, and disposal of sensitive information. Training should cover which data types require encryption, how to securely share files internally and externally (via the internet), and proper data retention and destruction procedures. When handling sensitive information, employees must understand their responsibilities under relevant industry regulations (such as GDPR, HIPAA, or PCI-DSS).

Mobile Device Security

After COVID-19, the mobile workforce has become a norm, and employees are increasingly using their devices for work-related business. Training should address securing both company-issued and personal devices used for work purposes. This includes proper device configuration, application permission controls, encryption requirements, and remote wipe capabilities. Employees should understand the risks associated with lost or stolen devices and procedures for immediately reporting such incidents to the IT department.

Reporting Procedures

A well-defined reporting procedure is important for handling potential security threats. Employees should be advised to report security incidents, suspicious activities, or vulnerabilities instantly to the organization's security operations center or IT security team without hesitation. This includes phishing attempts, lost or stolen computing devices, and unauthorized access attempts. A security culture that prioritizes prompt reporting without fear of blame ensures faster incident response and minimizes security risks. Clear escalation paths and contact information should be readily available to all employees.

Vulnerability and Patch Management

In the previous chapter, we discussed vulnerability and patch management, outlining a process for executing vulnerability management (VM). In this chapter, we will continue our discussion, focusing on implementing VM in an enterprise context.

It is important to clarify that the terms vulnerability, risk, and threat are often used interchangeably but have different meanings. Vulnerability refers to any weakness in one or more assets that a threat can exploit. For example, a software bug that malicious actors could target is a vulnerability.

On the other hand, a threat is any potential danger that could harm IT systems or data. A hacker or a malicious actor would be considered a threat.

Risk is the potential consequence that occurs if a threat exploits a vulnerability. For instance, there is a risk of a data breach if a threat actor exploits a software bug in a banking application that manages and processes critical customer data.

It is essential to remember that a vulnerability alone does not cause harm to IT systems and data unless it is exploited by a malicious actor (threat).

Ranking Vulnerabilities

Organizations rank vulnerabilities according to different criteria. A widely accepted ranking system in the cybersecurity industry is the Common Vulnerability Scoring System (CVSS).[10] CVSS is a free and open industry standard that uses numeric ratings to determine the severity of vulnerabilities. The National Vulnerability Database (NVD)[11] adds a severity rating for CVSS scores.

Vulnerability Management as a Service (VMaaS)

Having an in-house VM program is costly and requires continuous monitoring. For instance, if your company operates in highly regulated industries, such as healthcare or banking sectors, it is mandatory to deliver timely and continual reports about VM to remain compliant. Having a dedicated VM team could be outside most companies' budgets. To mitigate this problem, a new model in cloud computing has emerged, which provides VM services through a third-party provider. This allows your company to outsource VM tasks while leaving their staff to focus on more critical tasks and activities.

Hiring an external VM provider offers numerous benefits to companies, including:

- Cost-effective: VMaaS helps reduce the costs associated with maintaining an in-house VM program, including salaries, training, and infrastructure expenses.
- Better expertise and resources: VMaaS providers are specialized companies with extensive VM expertise and access to advanced tools that individual companies may find difficult to acquire.
- Scalability: VMaaS can quickly scale up or down based on your organization's needs, helping you cut operational costs and providing flexibility as your business grows or changes.
- Continuous monitoring and updates: VMaaS providers offer round-the-clock monitoring and are better equipped to stay updated with the latest vulnerability information through their connections to cyber threat intelligence feeds. This ensures that your IT systems are well-protected against emerging threats.
- Compliance management: VMaaS can help organizations meet regulatory requirements, such as those imposed by GDPR, PCI DSS, and HIPAA, by providing the necessary reports and documentation for audits.
- Integration capabilities: VMaaS solutions are designed to work with various IT environments, allowing them to integrate seamlessly with existing security tools and processes without the need for additional costly hardware or software.
- Comprehensive coverage: VMaaS typically covers both cloud-based and on-premises IT infrastructure, providing full visibility into vulnerabilities across your entire IT environment.
- Reduced workload on internal IT teams: By outsourcing VM activities, internal IT and security teams can focus on strategic tasks and core business functions, improving resource allocation across the organization's IT environment.

Vulnerability Scanning Tools

Various security products scan for vulnerabilities and recommend or adopt the appropriate patch for them. Here are the most prominent vulnerability scanning solutions:

- Acunetix[12]: An automated web vulnerability scanner that searches for & fixes vulnerabilities in websites, applications & APIs (commercial license)
- BeSECURE[13]: A vulnerability assessment and management tool compatible with different systems such as Microsoft, UNIX, Novell, and network devices.
- Burp Suite[14]: A popular web vulnerability scanner.
- Nessus[15]: A well-known vulnerability scanner tool. It can also scan cloud instances and provide compliance and security audits for compliance bodies.
- OpenVAS[16]: A free vulnerability scanner program.

A vulnerability scan should be performed on a monthly basis to avoid leaving open vulnerabilities in IT systems for a long period.

Security Audits

Security audit, also known as cybersecurity audit, is an important element in any cybersecurity defense strategy. IT security audits give organizations a clear picture of their organization's surrounding risk environment and how they are prepared to handle different types of threats against their digital assets.

IT security audit is a comprehensive assessment of an organization's security posture; it measures how well your security policies, technical controls, and procedures meet established security standards. An IT security audit covers two main security areas:

1. Technical aspects, such as firewall and IPS/IDS systems, in addition to application security and the physical components of your information system
2. The human side, such as employees' ability to detect social engineering attacks and how they collect, store, and process sensitive business information

Security audits employ various methodologies to uncover potential weaknesses across the IT landscape, including penetration testing, vulnerability scanning, and control assessment. Organizations should establish audit frequencies based on their industry, with high-risk sectors like banking and healthcare providers often requiring quarterly assessments, while others may execute bi-annual reviews.

By assessing both sides, organizations can implement the best security and managerial procedures to protect sensitive data and other digital assets, implement proper security controls, identify security vulnerabilities, and manage access controls across the entire organization's IT environment. The audit assessment conclusion typically produces detailed reporting documents highlighting discovered gaps and proposing remediation strategies with priority levels to address each weakness.

Emerging Audit Concerns

Emerging audit concerns now include cloud infrastructure vulnerabilities, remote work environment exposures, and Internet-of-Things (IoT) device risks, as these modern technological deployments create additional attack vectors beyond traditional network boundaries. For instance, organizations adopting these technologies will face expansive threat landscapes requiring comprehensive security assessments beyond perimeter-focused evaluations.

Cloud Infrastructure

Cloud infrastructure vulnerabilities appear when organizations shift sensitive business data to third-party platforms, such as AWS and Google Cloud, without implementing proper access controls. For instance, misconfigured Amazon Web Services (AWS) S3 buckets have exposed millions of customer records[17] when default permissions remained unchanged, which allows unauthenticated access to confidential information. In the same way, inadequate API security protocols create exploitable weaknesses when authentication mechanisms lack multi-factor verification requirements.

Remote Work

Remote work environment exposures arise from employees using personal devices and unsecured networks, such as home and public, for corporate work activities. Consider the finance department employee accessing sensitive financial records via public Wi-Fi in restaurants and bus stations while working remotely. This makes their connections susceptible to packet sniffing attacks. On the other hand, corporate VPN solutions deployed quickly during workforce transitions often lack sufficient endpoint security validation, which allows compromised devices to connect directly to internal resources.

Internet-of-Things

Integrating IoT devices into operational technology (OT) systems presents great security risks, as these devices bridge the gap between physical operations and digital information networks. These risks are intensified by utilizing common weak security practices such as using hardcoded credentials in IoT devices with limited firmware update capabilities and long device lifespans (such as digital door locks), which create persistent vulnerabilities in corporate environments.

For example, manufacturing facilities often deploy smart sensors to monitor production lines, optimize efficiency, and predict maintenance needs. However, many of these devices come with hardcoded usernames and passwords that cannot be changed, such as "admin/admin" or "root/1234". This makes them easy targets for attackers who can exploit these default credentials to gain unauthorized access to broader network places. Once inside, attackers can move laterally across the network to disrupt production processes or steal sensitive intellectual property. A notable real-world example is the Triton malware attack on a petrochemical plant in 2017,[18] where attackers exploited insecure IoT devices to gain access to safety systems. The attack led to a catastrophic failure.

In the same way, building management systems (BMS) used to control environmental systems (e.g., HVAC, lighting) and physical access (e.g., door locks, surveillance cameras) often remain in operation for decades with outdated firmware. These systems are rarely updated due to the complexity of managing legacy devices or the fear of disrupting critical operations (such as building access). For instance, in 2014, attackers breached a major retail chain's network by exploiting vulnerabilities in an HVAC system that had not been updated in years.[19] This breach led to the theft of 110 million customers' credit card information. The outdated firmware in these systems acts as an exploitable access vector, allowing attackers to pivot from seemingly low-risk devices to more critical corporate networks.

Is It Necessary to Conduct Security Audits?

It is worth noting that an IT security audit is mandatory for big companies – in most cases – as most compliance frameworks require regular security audits to assess an organization's ability against cyber threats. For example, financial services, like banks and credit unions, need to comply with the Sarbanes-Oxley Act (SOX) regulation, while healthcare providers need to comply with the Health Insurance Portability and Accountability Act (HIPAA). Credit card companies have their own framework, PCI DSS (Payment Card Industry Data Security Standard). In general, any company that processes or stores customers' personally identifiable information or patients' information must perform a security audit to maintain the highest security for its IT systems.

Organizations may utilize internal audit teams to conduct routine assessments. Still, specific compliance frameworks (e.g., ISO 27001 and COBIT) mandate engaging third-party auditors to ensure an unbiased examination of security measures without potential conflicts of interest.

Another reason to conduct regular security audits is dealing with government agencies or working in foreign markets that require regular audit assessments to ensure the security of customer data.

Incident Response and Recovery

Incident response is the set of processes and technologies used to detect and respond to cyberthreats and other forms of unexpected incidents. Incident Response (IR) is a part of the more comprehensive Incident Management, which also includes other non-technical departments of the organization, such as legal and public relations departments.

The aim of incident response is to prevent an incident from escalating and to prevent cyber incidents from causing system downtime when they occur.

Each organization will have its incident response plan (IRP). This plan identifies the types of incidents in addition to how to detect (identify) and resolve each one. A proper and well-organized IRP will allow an organization to respond swiftly to cyberattacks and restore operations as fast as possible after a cyberattack. The reduction of downtime will allow an organization to serve its clients better and maintain a strong security posture. All of these can be achieved by a proper IRP.

What Is a Security Incident?

A security incident, also called a security event, is any unexpected digital or physical error that could impact the confidentiality, integrity, or availability of an organization's IT systems or sensitive data. Under this definition, cyber incidents cover wide arrays of security events ranging from hackers infiltrating internal networks, unaware users installing malicious programs inadvertently on their devices, or wrong endpoint configurations that result in a data breach. Here are some common security incidents:

- **Ransomware attack**: Malicious software that encrypts systems and demands payment for decryption keys.
- **Phishing and other types of social engineering attacks such as smishing and quishing**: Deceptive attempts to steal credentials or install malware via fraudulent email, SMS, or in-person communications.
- **DDoS and DoS attacks**: Overwhelming systems with false traffic to render the online services unavailable.

- **Insider threats**: Malicious actions conducted by employees or third-party contractors with legitimate access to internal systems.
- **Privilege escalation attacks**: Unauthorized elevation of system access rights
- **Man in the middle attacks**: Intercepting communications between two parties without their knowledge. This commonly occurs when exchanging sensitive data over insecure connections.

Incident Response Plan

A typical IRP is composed of the following key elements.

Preparation

The first phase involves preparing for cyber threats or incidents that may occur to your IT systems and/or data. The preparation needs time to finish and usually occurs when there are no incidents.

It is essential to continually review this phase, as cyber threats are evolving rapidly. With each emerging threat, the preparation phase should be reviewed to incorporate the new threat or attack vector into consideration. For example, when utilizing AI solutions at work, it is essential to have a clear understanding of how to handle AI-related cyber threats, such as prompt injection attacks or AI-generated phishing content.

Detection

The detection phase involves detecting the threat as early as possible to prevent it from causing damage. Detection is commonly dependent on seeing the indicators of compromise and indicators of attacks to capture the security risks and issue alarms before they escalate within the target IT environment.

To counter the ever-increasing number of threats, detection commonly depends on leveraging automated solutions to scan large numbers of security alerts and detect which ones are actual threats that should be escalated to the incident response team. For instance, SIEM systems can correlate logs from multiple sources to identify suspicious patterns that might indicate an active intrusion. At the same time, EDR solutions can detect unusual endpoint behavior that might signal malware execution.

Analysis

After identifying the threat, the next step is to analyze this threat to understand who is behind it, how it was developed, and what its aim is from the attack. This allows incident response teams to prioritize threats based on their severity and impact, which also results in better optimizing available resources.

The analysis phase requires using digital forensics tools, automated tools, and human experience to produce a coherent result for each discovered event. For example, analyzing network traffic captures can reveal command and control communications used during ransomware attacks, while memory forensics can uncover rootkits or fileless malware that traditional antivirus may not detect.

Contain the Threat

It is worth noting that many threats can be contained before they cause any damage to target systems. When there is the right expertise and adequate security tools, then incident response teams can contain incidents and halt them efficiently.

For example, by isolating infected systems or shutting the system completely, incident response teams can prevent severe damage caused by a ransomware attack. However, incident response teams should always balance the ability to cease system operations with the damage if they leave critical systems compromised.

Eliminate the Threat

In this phase, the incident response team will remove all traces of the cyber attackers from the system.

It is important to scan all infected systems to ensure you are not leaving any system infected with malware. Everything from the point of infection until the eradication point should be inspected thoroughly to ensure the risk has been removed completely.

Tracing the attack back from its origin to its detection allows incident response teams to understand the root cause of the problem, which gaps or security vulnerabilities the attackers used to exploit the system. This allows them to close these gaps to prevent future attacks from the same spots.

Recovery

After eliminating the threat, systems need to be restored to their normal operation. This phase involves the following:

- Rebuilding affected systems from clean backups
- Restoring data from verified backup sources
- Verifying system integrity before returning to production
- Implementing additional monitoring for affected systems

Lessons Learned

The final phase involves conducting a thorough post-incident review to understand what happened during the incident, how it happened, and how the response could be improved in the future. This includes documenting the incident timeline, measuring the effectiveness of the response, and identifying areas for improvement.

During this phase, the incident response team should:

- Document every aspect of the incident and the response efforts
- Identify security gaps that contributed to the incident
- Update security controls and procedures based on their discovery
- Revise the IRP if necessary
- Provide additional cybersecurity training to staff based on lessons learned

For example, suppose the incident originated from a phishing email that bypassed email filters. In that case, the organization might need to update email security controls and conduct additional security awareness training for employees.

INCIDENT RESPONSE FRAMEWORKS

Some government and large organizations with significant security expertise have developed incident response frameworks to guide other organizations to adopt when managing cybersecurity and other types of incidents. Here are the most prominent ones:

1. SANS Incident Response Framework[20]
2. NIST Incident Response Framework[21]

Tools for Incident Response

The incident response team can utilize different types of tools during their incident response work. Here are the most prominent ones.

Security Orchestration, Automation, and Response

Security orchestration, automation, and response (SOAR) solution is a collection of tools that gather security data from various sources and present it to the security team in a unified dashboard. Current SOAR solutions utilize AI and ML technologies to automate threat detection and incident response activities. A SOAR system is composed of three main components.

Security Orchestration

This component connects and integrates different security tools from external and internal sources to work in harmony. These tools include:

- IDS and IPS
- Firewalls
- SIEM
- Vulnerability scanners
- EDR solutions
- Threat intelligence feeds
- End-point security solutions
- User and entity behavior analytics

For instance, when an EDR solution detects a potential malware, orchestration enables automatic enrichment of this alert with relevant data from threat intelligence platforms and user behavior analytics to determine the severity level.

Security Automation

Based on the collected data from the previous tools, the automated part executes an automated process to replace manual tasks. By leveraging AI and ML technologies, SOAR can be used to prioritize threats, make recommendations to handle them, and automate future responses. For example, upon detecting multiple failed login attempts from an unusual geolocation, a SOAR platform might automatically block the IP address, reset affected credentials, and create an investigation case without human intervention.

Security Response

After detecting the threat, SOAR executes relevant response procedures. It also provides reporting capabilities after responding to an attack. In practice, this might involve isolating compromised endpoint devices from the network, initiating forensic analysis, and generating comprehensive incident reports that highlight attack vectors and the remediation steps taken.

Security Information and Event Management

Security Information and Event Management (SIEM) is a cybersecurity management system that collects data from various endpoint devices and networking equipment and displays them in a single console. It allows security teams to monitor different network parameters to take actions to remediate threats before they pose risks to the target IT environment.

Modern SIEM solutions allow for the detection of any abnormal behaviors of user activities using AI and ML technologies. This allows automation of many tasks associated with threat detection and response.

SIEM tools were originally composed of log management tools; however, they have added other tools such as security information management (SIM) and security event management (SEM) functions, to provide a more unified view for security teams.

For instance, an SIEM might correlate failed login attempts across multiple systems with unusual network traffic patterns to identify a potential breach in progress. When suspicious activity is detected, the SIEM can trigger alerts based on predefined thresholds, such as multiple failed authentication attempts from foreign IP addresses within a short timeframe.

Threat Intelligence

Threat intelligence is the practice of collecting threat data from various sources to understand attackers' targets, motivations, techniques, and tools used to execute their attacks and act upon them. It converts raw data into actionable insights, enabling better decision-making. Threat intelligence provides actionable insight into current and future threats, such as emerging threats. It helps security teams to take one step ahead of cyber attackers by understanding their attack methods and techniques to halt their attack plans and boost their security defenses.

The rise of APT attacks necessitates the need for threat intelligence capability to understand adversaries' tactics, techniques, and procedures (TTPs). This knowledge will allow security teams to boost their cybersecurity defenses and follow a proactive approach to halt cyberattacks before they knock on organization doors.

Although threat intelligence has become widely known as a term among organizations, many companies are still leveraging it superficially. For instance, they utilize information collected from threat intelligence feeds into their installed security solutions, such as SIEM, Firewalls, IPS/IDS, and SOAR, and let the tools act automatically. While this approach is one side of using threat intelligence data, it is not enough. Threat intelligence has more capabilities to offer than this.

Threat Intelligence Capabilities

Threat intelligence can be used in the following cases:

- **Uncover unknown threats**: Such as APT and ransomware attacks and act before they cause severe harm to target organization data and systems. For example, detecting a specific command and control server pattern associated with an emerging threat actor targeting your industry sector.

- **Understand adversary behavior**: Knowing attackers' TTPs will allow defenders to understand their attack techniques and motivations and plan their defense accordingly. For example, recognizing that a particular threat actor group exploits unpatched VPN vulnerabilities as an initial access vector.

- **Provide information for decision-makers**: Decision makers in non-technical positions such as CEO and top management can base their decisions in different fields based on threat intelligence information. For example, they can plan their budget or decide to join a partnership with a vendor based on threat intelligence data.

- **Proactive defense**: Threat intelligence allows defenders to move from a reactive defense that waits for an incident to respond to it to a proactive defense that tries to prevent the incident from occurring in the first place. This might involve blocking specific IP ranges known to host malicious infrastructure before they attempt to breach your network.

What Organization Size Benefits from Threat Intelligence Information

Regardless of their size, any organization can benefit from threat intelligence data. For example:

- Medium-size organizations can use threat data from public and private sources to plan their defenses and see what threats will likely hit them, mostly according to their business sector or OT. A healthcare provider might focus on intelligence related to ransomware campaigns targeting medical records systems.

- Large organizations need threat intelligence to integrate external threat intelligence sources into their internal security efforts to have a holistic view of the threat landscape. A multinational financial institution might correlate global banking threat feeds with its security telemetry to identify targeted campaigns against its specific assets.

Threat Intelligence Lifecycle

A typical threat intelligence lifecycle is composed of the following six phases.

Requirements

In the first phase, we define the goals and methodology of the intelligence program that align with stakeholder needs. For example, a financial institution wants to protect its online banking platform from cyberattacks. The security team meets with stakeholders such as executive leadership, IT, and compliance departments to define the goals of the program.

Here are some questions to ask in the first phase:

- Who might target us? (e.g., financially motivated threat actors, nation-state actors such as North Korea and Russia).
- What are our potential vulnerabilities? (e.g., weak authentication, outdated software, legacy hardware systems, such as digital doors looks, that cannot accept new updates.
- What actions can we take to improve defenses? For example, implement MFA, change legacy systems or disconnect them from main IT system, patch systems, purchase and install new security solutions such as EDR on computing devices.

Collection

In the second phase, we gather information from various sources to address the defined requirements. For example:

The security team will collect threat data from:

- Internal sources: Firewall logs, IDS/IPS alerts, and EDR tools
- External sources, such as:
  - Threat feeds (e.g., reports on new banking Trojans like Emotet). There are also many threat intelligence feeds, and some are free such as SANS Internet Storm Center (ISC),[22] LevelBlue Labs Open Threat Exchange (OTX),[23] and Spamhaus Project.[24]
  - Dark web forums (e.g., monitoring for mentions of the organization's name, domain name, or leaked credentials) in darknets such as TOR and I2P.
  - Social media (e.g., identifying phishing links shared on social media platforms such as X and Facebook). You can also check phishing database repositories such as Phishing Database Project[25] and PhishTank.[26]
  - Open-source intelligence (OSINT) tools (e.g., Shodan to find exposed servers). We already covered OSINT in a previous chapter.

For example, the outcome of the collection phase could be gathering a list of IP addresses, malware samples, and phishing URLs associated with recent attacks on financial institutions.

Processing

In the processing phase, the intelligence team will organize and clean raw data into a format suitable for analysis. For example, the team processes the collected data using the following methods:

Decrypting files: Unpacking malware samples in a sandbox environment to analyze their behavior. Any.run[27] and Joesandbox[28] are examples of free sandboxes to analyze malware.

Translating collected data: Converting foreign-language discussion forum posts (e.g., Russian, Arabic, or Chinese) into English using translation tools. Here are three free translation services: Google Translate,[29] DeepL,[30] and Microsoft Translate[31]

Formatting data: Structuring IP addresses, domains, and file hashes into a spreadsheet for easier analysis. You can use LibreOffice Calc[32] and Zoho Sheet[33] as alternatives to MS Excel application

After organizing all collected data and grouping them into categories, the final outcome of the processing phase is to have a structured dataset of indicators of compromise (IOCs), such as malicious IPs, domains, and file hashes.

Analysis

In the analysis phase, the processed data is analyzed to produce actionable insights and recommendations for concerned stakeholders.

For example, the team analyzes the data to answer the questions from the requirements phase:

Who is targeting us? For example, suppose a bank was hit with a large number of suspicious emails. After inspecting the attachment within these emails, the security teams discovered it was containing the Emotet malware (after inspecting the attachments), which is known to be operated by a threat actor group called "TA505", which widely uses the Emotet malware to attack financial institutions. The threat actor was identified after analyzing attachments sent in numerous email messages. The attachments contained the Emotet malware after inspecting the program signature.

What are their tactics? The group sends phishing emails with malicious attachments to bank employees. For example, the attackers send phishing emails disguised as legitimate communications from a trusted third-party vendor or a government body. For instance, an email might appear to be from the "Federal Reserve Bank" with a subject line like "Urgent: Compliance Review". The email contains a malicious Word document attachment that, when opened, installs Emotet on the victim's machine.

What are our vulnerabilities? The bank's email filtering system fails to detect phishing emails because the attackers are using newly registered domains (NRDs) that are not yet flagged as malicious, and the email filtering system relies solely on static blacklists. On the other hand, some employees are not trained to recognize phishing attempts and inadvertently open malicious attachments after executing a simulated phishing campaign.

What actions can we take? The security team may recommend the following:

- Updating email filters to block the identified phishing domains. They also implement a dynamic filtering solution that uses threat intelligence feeds to automatically block newly reported malicious domains.
- Training employees to recognize phishing attempts.
- Blocking the malicious IPs and domains. After identifying the Emotet command-and-control (C2) servers and their IP addresses, the security team blocks them at the network level (Firewall).

The final outcome of this phase is that the team produces a report detailing the threat actor's TTPs and provides actionable recommendations to prevent future attacks.

Dissemination

In this phase, the security team presents findings in a digestible format tailored to the stakeholder audience. For example, if the stockholder is top management and is not tech-savvy, then our report should describe technical jargon in an easy-to-digest language for them.

In the dissemination phase, the team prepares two versions of the report:

- Technical report: This report is directed to IT and security teams and includes detailed IOCs, malware analysis, and recommended mitigation steps.
- Executive summary: This report is directed to management leadership, focusing on the business impact (e.g., potential financial losses) and high-level recommendations (e.g., investing in employee training).

The final outcome of this phase is that the IT team implements the recommended security measures while leadership approves a budget for additional training and/or the purchase of tools.

Feedback

In this phase, we gather feedback from different stockholders to refine future threat intelligence operations. For example, after implementing the recommendations, the security team meets with stakeholders to gather feedback:

- IT team: Reports that the new email filters reduced phishing emails by 90%
- Leadership: Requests more frequent updates on emerging threats, such as APT and ransomware attacks
- Compliance team: Suggests including regulatory implications in future reports.

The final outcome should adjust work processes to meet stockholders' expectations:

1. Increases the frequency of threat intelligence updates
2. Adds a section on compliance risks to future reports
3. Prioritizes monitoring for new phishing campaigns

AI and Machine Learning in Cybersecurity

As cyber threats grow in sophistication and number, traditional security solutions are no longer enough to keep pace. However, AI and ML have come into play to change this. For instance, these modern technologies are revolutionizing the way we detect, prevent, and respond to cyber threats. By leveraging AI and ML, cybersecurity professionals can analyze vast volumes of data (such as network traffic and information collected from public sources), identify suspicious patterns, and predict potential attacks quickly and accurately.

AI and ML are transforming cybersecurity by enabling proactive and adaptive defense mechanisms. One of their most notable contributions is in threat detection and prevention. Traditional security systems rely on predefined rules and signatures to identify threats; this makes them ineffective against new or evolving attacks. In contrast, AI and ML algorithms analyze historical and real-time data to detect anomalies and suspicious patterns that may indicate a cyber threat. For example, ML models can identify zero-day vulnerabilities more efficiently by recognizing unusual behavior in network traffic or system activity. This capability is particularly valuable in combating APT attacks, where attackers try to remain hidden for extended periods.

Another critical application is the automation of security processes. Cybersecurity teams are often overwhelmed by the sheer volume of alerts and tasks, such as log analysis, vulnerability scanning, and patch management. AI-powered tools can automate these repetitive tasks, which helps free up human analysts to focus on more complex issues. For instance, Google Security Operations[34] uses ML to analyze billions of security events daily, assisting organizations to prioritize and respond to threats more efficiently.

AI and ML also excel in behavioral analysis. For example, by monitoring user and entity behavior when interacting across the organization IT environment, these technologies can identify deviations from normal usage patterns that may indicate malicious activity. For example, if an employee's account in the marketing department suddenly starts accessing sensitive files at unusual hours in the finance department, an AI system can flag this as a potential insider threat. In the same way, ML algorithms can detect phishing attempts by analyzing email content and sender behavior, even if the attack is highly targeted.

Beyond these applications, AI-powered solutions are demonstrating remarkable effectiveness in malware detection. Traditional antivirus solutions struggle against polymorphic malware that constantly changes its code to evade detection. Cylance (now acquired by Arctic Wolf),[35] for instance, utilizes ML algorithms to analyze millions of file characteristics to identify malicious software before it executes and can achieve a higher detection rate for previously unseen malware variants.

AI systems can also enhance incident response capabilities. For instance, it provides:

- Faster detection through anomaly detection and real-time analysis: The ability to analyze vast amounts of data in real-time allows ML models to compare normal data traffic with abnormal activities to flag suspicious behaviors instantly.
- Improved incident analysis via pattern recognition and threat intelligence: AI-powered tools can analyze historical data and identify traffic patterns associated with known threats. These solutions can also integrate with threat intelligence feeds to correlate current incidents with known TTPs.
- Automate incident prioritization: AI can automatically prioritize incidents based on their severity, potential impact, and success rate. This reduces the workload of human analysts and ensures that the most critical incidents are handled first.

Practical Applications of AI and ML in Cybersecurity

The applications of AI and ML in cybersecurity are vast and varied and can strengthen current security solutions to combat evolving threats. Below are the most prominent areas where these technologies are making a substantial impact.

Malware Detection

Traditional antivirus software relies on signature-based detection, which struggles to identify new or polymorphic malware. In contrast, ML models can analyze the behavior of files and processes to detect malicious activity within the entire IT environment. For example, Microsoft's Windows Defender uses ML to identify and block malware in real-time, even if it has never been encountered before. This approach is particularly effective against emerging threats, such as zero-day threats, where attackers exploit vulnerabilities before their software and hardware vendors patch them.

Phishing and Fraud Detection

AI-powered solutions excel at identifying phishing attempts by analyzing email content, URLs, and user behavior. For instance, Google's Gmail uses ML to block more than 99.9%[36] of spam, phishing, and malware every day. It can also flag suspicious links and attachments before they reach users. In the same way, financial institutions leverage AI to detect fraudulent transactions. Companies like PayPal use ML algorithms to analyze spending patterns and flag abnormal behaviors, such as unusual purchase locations or spending too much. This helps prevent millions in potential fraud losses each year.

Network Security

AI and ML are transforming network security by enabling real-time monitoring and intrusion detection. For instance, many cybersecurity solutions providers use ML to create a "self-learning" model of normal network behavior. Any deviation from this – normal – model, such as unusual data transfers or unauthorized access attempts, is flagged as a potential threat. This allows organizations to respond quickly to attacks, such as ransomware or data exfiltration attempts.

Endpoint Protection

With the rise of remote work after the COVID-19 pandemic, securing mobile computing devices like laptops and smartphones has become more challenging. AI-powered endpoint protection solutions can predict and prevent attacks by analyzing device behavior and identifying suspicious activity. For example, AI-powered solutions use ML to detect and block threats on endpoints in real-time, such as unauthorized access attempts or running malicious processes. In the same way, autonomous endpoint protection solutions leverage AI to automatically respond to threats, such as isolating compromised devices or rolling back ransomware attacks to prevent distributing malware across the entire IT environment.

Challenges and Limitations

Despite their huge potential, AI and ML in cybersecurity come with challenges. One major issue is data quality. AI systems rely on large amounts of high-quality data to train the underlying ML models. However, many organizations struggle to collect and maintain clean, relevant data. For instance, financial institutions face difficulties generating training data sets that accurately represent evolving money laundering schemes without including legitimate transactions. On the other hand, cybercriminals are increasingly using adversarial ML to bypass controls set by AI systems to prevent them from producing malicious output.

Another challenge is the risk of false positives and negatives. While AI can significantly improve threat detection, it is not infallible. Over-reliance on AI can lead to missed threats or unnecessary alerts, which result in overwhelming security teams with false alerts. For example, a major retailer's AI-based threat detection system might flag hundreds of legitimate customer access attempts as suspicious during peak shopping seasons due to unusual but normal traffic patterns. Finally, there are ethical and privacy concerns. The use of AI in surveillance and data analysis raises questions about user privacy and the potential for misuse.

Another significant limitation is the skills gap in the cybersecurity workforce. Organizations often lack personnel with expertise in both cybersecurity and data science, making it difficult to develop, deploy, and maintain effective AI-driven security solutions. For instance, a manufacturing company implementing anomaly detection across its OT environment might struggle to find specialists who understand both industrial control systems and ML algorithms necessary for proper implementation.

Summary

In today's information age, organizations rely heavily on IT systems, which makes having a robust cybersecurity strategy essential to safeguard digital assets and ensure business continuity. The cost of inadequate defenses can be catastrophic, with data breaches averaging $4.45 million in 2023. A strong cybersecurity strategy not only mitigates financial losses but also enhances competitive advantage and stakeholder trust in addition to maintaining organization reputation.

This chapter focuses on enterprise network defense strategies. It emphasizes the importance of securing the network perimeter – the first line of defense against external threats. The key components of perimeter security include firewalls, IDS/IPS, border routers, and VPNs. Firewalls, for instance, come in various forms – hardware, software, SaaS, or cloud-based – and types like a proxy, stateful inspection, and NGFWs, each offering unique security features as we saw. NGFWs, for example, provide advanced capabilities like SSL/TLS inspection and integration with threat intelligence feeds.

Network segmentation is another critical strategy. It works by dividing the network into isolated segments to limit the spread of threats and improve performance. This approach aligns with ZTA, which enforces strict access controls and continuous verification across the entire IT environment and ensures no user or device is trusted by default.

Penetration testing and physical security measures are vital for identifying vulnerabilities and protecting IT infrastructure from physical threats like theft or natural disasters. Security policies and procedures, including access controls, incident response, and employee training, further strengthen defenses by addressing both technical and human factors.

Finally, AI and ML are transforming cybersecurity by enabling faster threat detection, automated incident response, and advanced behavioral analysis. These technologies help organizations stay ahead of evolving threats, ensuring a proactive and adaptive defense posture.

In summary, a comprehensive cybersecurity strategy must combine robust technical controls, proactive testing, and employee awareness to protect enterprise networks from increasingly sophisticated threats.

Notes

1   Pcisecuritystandards, "PCI DSS Quick Reference Guide", Accessed 2025-03-28. https://listings.pcisecuritystandards.org/documents/PCI_DSS-QRG-v3_2_1. pdf
2   Snort, "Snort Tool", Accessed 2025-03-28. https://www.snort.org
3   Ossec, "OSSEC HIDS", Accessed 2025-03-28. https://www.ossec.net
4   Suricata, "Suricata Tool", Accessed 2025-03-28. https://suricata.io
5   Google cloud, "BeyondCorp", Accessed 2025-03-28. https://cloud.google.com/beyondcorp
6   Microsoft, "Zero-Trust", Accessed 2025-03-28. https://www.microsoft.com/en-us/security/business/zero-trust
7   IBM, "Data Breach", Accessed 2025-03-28. https://www.ibm.com/reports/data-breach
8   Authenitc8, "Business Email Compromise", Accessed 2025-03-28. https://authentic8.com/blog/business-email-compromise-phishing
9   Authenitc8, "Types of Phishing Attacks", Accessed 2025-03-28. https://authentic8.com/blog/types-of-phishing-attacks
10 First, "Common Vulnerability Scoring System SIG", Accessed 2025-03-28. https://www.first.org/cvss/
11 NIST, "National Vulnerability Database", Accessed 2025-03-28. https://nvd.nist.gov
12 Acunetix, "Acunetix Tool", Accessed 2025-03-28. https://www.acunetix.com
13 Fortra, "Fortra Vulnerability Management", Accessed 2025-03-28. https://www.beyondsecurity.com/products/besecure

14 Portswigger, "Portswigger Solution", Accessed 2025-03-28. https://portswigger.net/burp/enterprise

15 Tenable, "Nessus", Accessed 2025-03-28. https://www.tenable.com/products/nessus

16 OpenVAS, "OpenVas Tool", Accessed 2025-03-28. https://www.openvas.org

17 Awsinsider, "Yet Another Misconfigured AWS S3 Bucket Exposes Sensitive Customer Data", Accessed 2025-03-28. https://awsinsider.net/Articles/2024/12/10/Yet-Another-Misconfigured-AWS-S3-Bucket-Exposes-Sensitive-Customer-Data.aspx

18 Technologyreview, "Triton is the World's Most Murderous Malware, and It's Spreading", Accessed 2025-03-28. https://www.technologyreview.com/2019/03/05/103328/cybersecurity-critical-infrastructure-triton-malware

19 Senate, "A Kill Chain Analysis of the 2013 Target Data breach", Accessed 2025-03-28. https://www.commerce.senate.gov/services/files/24d3c229-4f2f-405d-b8db-a3a67f183883

20 SANS, "Incident Response", Accessed 2025-03-28. https://www.sans.org/security-resources/glossary-of-terms/incident-response

21 NIST, "Incident Response", Accessed 2025-03-28. https://csrc.nist.gov/projects/incident-response

22 SANS, "Internet Storm Center", Accessed 2025-03-28. https://isc.sans.edu

23 Alienvault, "OTX", Accessed 2025-03-28. https://otx.alienvault.com

24 Spamhaus, "Spamhaus Project", Accessed 2025-03-28. https://www.spamhaus.org

25 Github, "Phishing Database", Accessed 2025-03-28. https://github.com/Phishing-Database/Phishing.Database

26 PhishTank, "PhishTank", Accessed 2025-03-28. https://phishtank.org

27 Any Run, "any.run", Accessed 2025-03-28. https://any.run

28 Joesandbox, "Joe Sandbox Cloud Basic", Accessed 2025-03-28. https://www.joesandbox.com

29 Google, "Googel Transdlate Service", Accessed 2025-04-02. https://translate.google.com

30 Deepl, "Translator", Accessed 2025-03-28. https://www.deepl.com/en/translator

31 Microsoft, "Translate", Accessed 2025-03-28. https://translator.microsoft.com

32 Libreoffice, "Calc", Accessed 2025-03-28. https://www.libreoffice.org/discover/calc

33 Zoho, "Sheet", Accessed 2025-03-28. https://www.zoho.com/sheet

34 Google, "The Business Value of Google Security Operations", Accessed 2025-03-28. https://services.google.com/fh/files/misc/idc-business-value-of-google-secops.pdf

35 Cylance, "The Arctic Wolf", Accessed 2025-03-28. https://arcticwolf.com/cylance

36 Google, "Protecting Businesses Against Cyber Threats During COVID-19 and Beyond", Accessed 2025-03-28. https://cloud.google.com/blog/products/identity-security/protecting-against-cyber-threats-during-covid-19-and-beyond

Index

For Product Safety Concerns and Information please contact our EU
representative GPSR@taylorandfrancis.com
Taylor & Francis Verlag GmbH, Kaufingerstraße 24, 80331 München, Germany